UX FUNDAMENTALS FOR NON-UX PROFESSIONALS

USER EXPERIENCE PRINCIPLES FOR MANAGERS, WRITERS, DESIGNERS, AND DEVELOPERS

Edward Stull

Apress®

UX Fundamentals for Non-UX Professionals: User Experience Principles for Managers, Writers, Designers, and Developers

Edward Stull
Upper Arlington, Ohio, USA

ISBN-13 (pbk): 978-1-4842-3810-3   ISBN-13 (electronic): 978-1-4842-3811-0
https://doi.org/10.1007/978-1-4842-3811-0

Library of Congress Control Number: 2018956323

Managing Director, Apress Media LLC: Welmoed Spahr
Acquisitions Editor: Shiva Ramachandran
Development Editor: Laura Berendson
Coordinating Editor: Rita Fernando

Cover designed by eStudioCalamar

Distributed to the book trade worldwide by Springer Science+Business Media New York, 233 Spring Street, 6th Floor, New York, NY 10013. Phone 1-800-SPRINGER, fax (201) 348-4505, e-mail orders-ny@springer-sbm.com, or visit www.springeronline.com. Apress Media, LLC is a California LLC and the sole member (owner) is Springer Science + Business Media Finance Inc (SSBM Finance Inc). SSBM Finance Inc is a **Delaware** corporation.

For information on translations, please e-mail rights@apress.com, or visit http://www.apress.com/rights-permissions.

Apress titles may be purchased in bulk for academic, corporate, or promotional use. eBook versions and licenses are also available for most titles. For more information, reference our Print and eBook Bulk Sales web page at http://www.apress.com/bulk-sales.

Any source code or other supplementary material referenced by the author in this book is available to readers on GitHub via the book's product page, located at www.apress.com/9781484238103. For more detailed information, please visit http://www.apress.com/source-code.

Printed on acid-free paper

This book is dedicated to James F. Crews.

Contents

About the Author. vii

Acknowledgments ix

Introduction xi

Part I: UX Principles. I

Chapter 1:   UX Is Unavoidable. 3

Chapter 2:   You Are Not the User 9

Chapter 3:   You Compete with Everything 13

Chapter 4:   The User Is on a Journey 19

Chapter 5:   Keep It Simple 27

Chapter 6:   Users Collect Experiences 35

Chapter 7:   Speak the User's Language. 39

Chapter 8:   Favor the Familiar 43

Chapter 9:   Stability, Reliability, and Security .49

Chapter 10: Speed 57

Chapter 11: Usefulness. 63

Chapter 12: The Lives in Front of Interfaces 67

Part II: Being Human. 69

Chapter 13: Perception. 73

Chapter 14: Attention 89

Chapter 15: Flow. 95

Chapter 16: Laziness. 101

Chapter 17: Memory 105

Chapter 18: Rationalization 113

Chapter 19: Accessibility 117

Chapter 20: Storytelling. 123

Part III: Persuasion **131**

Chapter 21: Empathy 135

Chapter 22: Authority 147

Chapter 23: Motivation. 153

Chapter 24: Relevancy 157

Chapter 25: Reciprocity 165

Chapter 26: Product. 171

Chapter 27: Price 181

Chapter 28: Promotion 195

Chapter 29: Place 203

Part IV: Process **207**

Chapter 30: Waterfall, Agile, and Lean 209

Chapter 31: Problem Statements 221

Chapter 32: The Three Searches. 227

Chapter 33: Quantitative Research. 233

Chapter 34: Calculator Research. 243

Chapter 35: Qualitative Research 247

Chapter 36: Reconciliation. 259

Chapter 37: Documentation 265

Chapter 38: Personas 277

Chapter 39: Journey Mapping 283

Chapter 40: Knowledge Mapping. 293

Chapter 41: Kano Modeling 299

Chapter 42: Heuristic Review 307

Chapter 43: User Testing 311

Chapter 44: Evaluation. 319

Chapter 45: Conclusion 323

Appendix A: Resources for Further Reading. 325

Index 341

About the Author

Edward Stull is a designer and researcher in Columbus, Ohio, USA. He helps teams work through user experience (UX) challenges ranging from product design to digital marketing. He has assisted numerous international brands, national banks, and state healthcare exchanges. He thinks a lot about how people understand, practice, and sell UX.

Acknowledgments

This book would not have been possible without the support of my family, friends, and colleagues. I would like to give special thanks to the following people for their assistance, advice, and encouragement: Robert Abbott, Arsalan Ahmed, Chase Banachowski, Laura Berendson, Robert Beuligmann, Casey Boyer, Brett Yancy Collins, Bill Crews, Dan Crews, Amanda Dempsey, Anthony Dempsey, Rita Fernando, Chris Flinders, Mark Gale, Bob Hale, Christopher Heidger, Sean Johnson, Brent Kaufman, Joe Kirschling, Neil Kulas, Bill Litfin, Jan Maroscher, Roy Nalazek, Stefanie Parkinson, Shiva Ramachandran, Paul Reiher, Nick Rieman, Charles Schmidt, Alberto Scirocco, David Scirocco, Marina Scirocco, Todd Sexton, Jim Smith, Carol Crews Stull, Edward L. Stull, Lalena Stull, Steve Swanson, Marty Vian, Adam Weis, and Emma Young.

Introduction

Every day, people like you and me experience the world. We wake up, pour a cup of coffee, and try to avoid stepping on the cat. Our day has just begun. Thousands of experiences await us, ranging from the trivial to the time-consuming: we peek at our phones and note our busy schedules, or we close our eyes and imagine a long, relaxing vacation under the sun. Some experiences are good. Some are bad. Most are somewhere in between. Yet, despite the many experiences we have, we are often unprepared to design new ones.

This is a book about designing experiences, a practice that goes as far back in time as human beings do. From ancient cave paintings to online wedding registries, people have designed experiences for millennia (see Figure I-1). Each time we find ourselves in a similar circumstance; wishing to create meaning from what we observe in the world and translate it for others. This is true whether we paint the wall of a cave or add a button to a website. We choose what. We select where. We determine how. If we do our job well, other people will understand the meaning we create: we will design their experiences.

Figure I-1. Wall painting in the Lascaux Cave.[1]

People of all ages design experiences. Young children host elaborate tea parties and build magnificent forts. Adolescents weave elaborate tales of love and struggle via text messages. Later in life, our experiences transform us into who we are—the thrills, the traumas, the grueling boredoms, the sweet seconds, and humbled hours. We become a collection of experiences.

Every culture designs its own experiences. Whereas a Japanese person may use both hands to present a business card during a formal exchange, presenting an item with your left hand is considered an insult in many Muslim countries. Deep within the forests of South Sulawesi, Indonesia, the Torajan people bury their deceased children inside the hollows of living trees, imparting a child's spirit into the leaves and branches above. American senators ascend their chamber's aisles to raid candy stockpiled within a designated mahogany desk[2]. Our culture shapes our experiences, and our experiences shape our culture.

Within every occupation, people design experiences. Millions of architects, engineers, playwrights, painters, bricklayers, and teachers fill our world with designed experiences. Onlookers gasp in wonder at the Burj Khalifa skyscraper, which soars 2,722 feet tall over the city of Dubai, while pearl divers on its nearby coast descend more than 100 feet on a single breath (see Figure I-2). State-sponsored commenters wade through millions of Internet posts website-by-website, negating criticisms of the Chinese government, much in the same way in that, each year, Girl Scouts sell millions of cookies door-to-door, negating Americans' diet plans. Like an endless assembly line, the working world creates, packages, and ships countless experiences each day.

---

[2]"Senate Chamber Desks." U.S. Senate: Select Committee on Presidential Campaign Activities. Accessed June 07, 2018. https://www.senate.gov/artandhistory/art/special/Desks/hdetail.cfm?id=1.

Figure I-2. Burj Khalifa skyscraper in Dubai.[3]

Regardless of age, culture, or occupation, much of what a person experiences is designed—be it a make-believe fort, an Indonesian funeral, or a box of Girl Scout cookies.

Increasingly, the experiences we design are digital. From apps to websites, from emails to video games, often the sole evidence of an experience appears on an illuminated screen. We create tiny worlds that thrive or perish at the whim of a device's on/off button. We make choices when we design, and based on these choices, our work shines in the daylight or declines into the recesses. The practice of user experience (UX) helps illuminate this uncertain terrain.

[3]Ashim D'Silva. "Burj Khalifa, United Arab Emirates." Digital image. Unsplash. April 4, 2006. Accessed June 6, 2018. unsplash.com/photos/rvyiu5qjI2E.

User experience fascinates me, but I do not assume you feel the same way. UX books tend to be written by and for UX designers. Few are written to include other roles: project managers, copy editors, graphic designers, and the like. This book is for everyone who works on digital projects. I wrote it with these people in mind. Perhaps this is you. This book seeks to inform and entertain, showing how UX has influenced history as well as our daily lives.

Rather than demonstrate concepts through a barrage of facts and figures, we will learn through stories. Poisonous blowfish, Russian playwrights, tiny angels, Texas sharpshooters, and wilderness wildfires all make an appearance. From 19th century rail workers to UFOs, we will cover a lot of territory, because the experiences that surround us are as broad and varied as every age, culture, and occupation.

What can a massive, WWII-era tank teach us about design? What does a small, blue flower tell us about audiences? What do drunk, French marathoners show us about software?

We start by covering the principles of UX. Next, we move on to being human. Afterward, we delve into a detailed discussion of persuasion. Finally, we wade into the murky waters of process. We talk about how you can navigate through all sorts of projects, what often works and what often does not, and why no philosophy is correct 100% of the time—including that which is found in this book.

By the time you reach the book's last pages, I hope you will find yourself somewhat changed, discovering new meaning and enjoyment in the experiences you create. This goal is simultaneously selfish and charitable. After all, we sometimes serve as the creators of an experience, but we are far more often the users of one. Experiences have always been this way. They always will be. Therefore, we can celebrate that UX not only enhances the world we now share, but also shapes the future yet to come.

UX Principles

If we were to believe the tale, the Lernean Hydra was a veritable horror show of fangs and fury. Nine heads sat atop its massive serpentine body. Breathing poison and snapping jaws, the hydra's heads would work in unison to simultaneously attack and defend against any would-be champions. Towering above any mortal, the mythological beast lived in isolation, because only the truly foolish would venture out and try to tame it.

When Hercules fought the hydra in the brackish swamps near Lake Lerna, the Greek hero had a few advantages (see Figure I-1). Though mortal, Hercules was favored by the gods. He had already defeated the ferocious, fabled Nemean lion and wore its impenetrable pelt like a suit of armor. Along with his legendary strength, Hercules was well on his way to BCE stardom. He had fought giants, mercenaries, and a virtual pan-Hellenic petting zoo full of other creatures. But the hydra was different. For each time Hercules would cut off one of its heads, two heads would grow back in replacement.

Figure I-1. Hercules slaying the Hydra, from The Labours of Hercules[1]

---

[1]Cropped version, Hans Sebald Beham "Hercules slaying the Hydra" from the *Labours of Hercules* (1542-1548).

Hercules would come to defeat the hydra by tackling one head at a time. He and his trusty assistant would lop off a head, cauterize the respective wound, and repeat the process until the job was done.

Defining UX principles can be a bit like battling a hydra. You tackle one principle, wait a short while, and two or more additional principles pop up in its place. It is a never-ending battle. Intriguing blog posts, inspiring speeches, and contentious twitter spats reshape our understanding of UX on a near-daily basis. However, some principles do endure.

This part of the book takes on the Herculean—and perhaps foolhardy—task of defining a core set of UX principles. The list is by no means exhaustive. A quick Google search of "UX principles" will return a long list of complementary approaches. As such, the following principles were selected to represent the enduring concepts shared among many approaches to user experience design and research.

User experience can first appear to be a big, scary monster of rules, contradictions, and dilemmas. While partially true, it is a monster easily tamed. We tackle one principle at a time, sear it into our memories, and become heroes to the users of what we create.

CHAPTER 1

UX Is Unavoidable

Palmolive released a series of TV commercials in 1981 featuring Madge, a spry and chatty manicurist. Each commercial's concept was simple: a housewife would visit a nail salon, inexplicably stick her hand into a small saucer full of green goo, then be told by Madge that the green goo was Palmolive dishwashing detergent. Surprise! By today's standards, the commercials were certainly gender-biased, if not borderline sociopathic, as Madge seemed to take great pleasure in telling unsuspecting housewives her trademark phrase: "You're soaking in it!"[1] Coined by the advertising firm Ted Bates & Co., the TV campaign reached legendary status by running continuously for nearly three decades. The campaign showed the power of a catchphrase and demonstrated a fundamental truth: we often do not realize our current circumstance until someone points it out to us.

The green goo we are soaking in today is user experience (UX), though you may not yet realize it. You feel it when you use products or services. You hear it in debates about features and functionality. You see its result when a project succeeds or fails. Like the unsuspecting housewife, you may not know what the green goo is; however, you still have your hand deep in the saucer. You cannot avoid UX—you may do it well or do it poorly, but either way, you are soaking in it.

---

[1] *Palmolive - "You're Soaking In It"*. YouTube. Accessed June 7, 2018. https://www.youtube.com/watch?v=dzmTtusvjR4.

© Edward Stull 2018
E. Stull, *UX Fundamentals for Non-UX Professionals*,
https://doi.org/10.1007/978-1-4842-3811-0_1

UX results from using any product or service. If you accept this premise, you will soon recognize the benefits of doing UX intentionally. Intentional user experience, or more precisely, *user experience research and design*, illuminates the needs of your audiences and creates compelling products and services. Conversely, unintentional user experience, or to put it more succinctly, *an accident*, foreshadows why audiences abandon and why products fail.

What Is User Experience?

The topic of user experience can bewilder people. The term *user experience* is itself somewhat confusing. It sounds simultaneously hippie and corporate, like a Grateful Dead poster affixed to the wall of an office cubicle.

The word "user" is the nominal form of "to use," which originates from the Old Latin verb "oeti," meaning "to employ, exercise, perform." The word "experience" originates from the Latin noun "*experientia*," meaning, "knowledge gained from repeated trials." Putting this all together, we arrive at the rough description of user experience to mean, "knowledge gained by doing something."

Don Norman, cofounder and principal at the Nielsen Norman Group, coined the term "user experience" decades ago. The term is [2]remarkably hardy, despite its occasional misinterpretation.

The umbrella term "user experience" covers several broad activities as the UX field continues to evolve. The field already includes aspects of cultural anthropology, human-computer interaction, engineering, journalism, psychology, and graphic design (many of which are terms not generally well understood by the public, either). These activities typically fall into one of two camps: the first is user experience design (UXD); the second is user experience research (UXR).

User experience design involves the design of a thing. That thing may be a product or service, or just a part of a product or service. For example, someone might design a web application to manage a nail salon, or design an iPhone app to file complaints about wayward manicurists.

User experience research includes primary research (i.e., discovery of original data), such as interviewing nail salon customers. In addition, it encompasses secondary, third-party research (i.e., reviews of previously discovered data), such as reading reports about customer behavior within the health and beauty sector.

---

[2]Norman, Don. "Peter in Conversation with Don Norman About UX & Innovation." Interview by Peter Merholz. Adaptive Path. December 13, 2007. Accessed June 7, 2018. http://adaptivepath.org/ideas/e000862/.

Much of what a UX professional does during her or his workday is dependent upon the mix of UXD and UXR required. Some firms have dedicated design and research roles, although many positions are often a combination of the two.

The Role of UX

Looking back over my years spent working within advertising and product design, I recall several times when a new colleague would walk into my office, sit in a chair, smile, and say, "So, what is it that you UX people do... exactly?" The question was often followed by a laugh, a deadeye stare, and the statement: "Really, I have no idea." Truly, many people have no idea what UX offers them.

The term "user experience" is not yet in the public's lexicon. Compounding the unfamiliarity are the many paths one might take to practicing user experience. Someone with a UX role may have a library sciences education, an engineering degree, formal training in psychology, or come from any number of other backgrounds. The variation complicates descriptions of UX roles outside the practice, as well as within it.

The Focus on Users

Because the field of user experience is broad, it is difficult to make many generalizations about UX roles. The commonality among all user experience roles is a focus on users. After all, user is in the name. Users, for lack of a better description, are people who use a product or service. You might think, "Well, my role considers such people. Why are UX roles even necessary?" I am glad you asked.

Various members of a project team set unique goals to reach. Account executives wish to reach client goals; managers, budgetary goals; strategists, strategic goals; visual designers, aesthetic goals; developers, technology goals. We rightly value these pursuits. Each is vital, as none is more or less important. However, we often forget why we perform these roles at all. We work for many reasons, but we ultimately work for the people who use what we create—we work for the users.

A UX practitioner aligns, refines, and reconciles business goals with what a user needs. Where business goals and user needs converge should be the sole determiner of functionality (see Figure 1-1). Build where they meet. Too often, project teams create features that address only business requirements, thereby neglecting user needs. Likewise, an application that provides only benefits to users erodes the underlying viability of the business that created it. After all, the motivation to produce an application is rarely altruistic. Even a charity wants its users to do *something*. So, the crucial question becomes, "How can we create experiences that address both user and business needs at the same time?" Let's consider the following example.

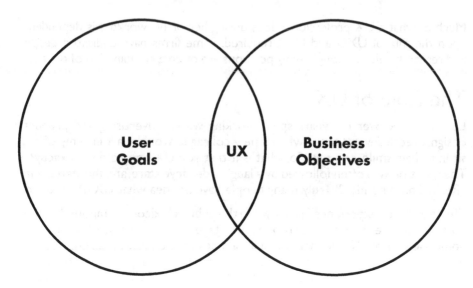

Figure 1-1. The convergence of user goals and business objectives

Our business goals are as follows:

- A high-end online beauty supply business wants customers to buy more products per visit.

- Being high-end, the business dislikes overt discounts and conspicuous promotions.

- The business wants to keep its current technology platform. The website's checkout process is awkward, but the current technology platform prohibits substantial modifications.

User needs are as follows:

- A user needs a competitive price to buy a particular product.

- The user also needs an efficient and easy-to-use way to make repeat purchases of the product in the future.

At first glance, we can see two issues requiring reconciliation. Increasing the number of purchases without providing discounts or promotions can be tough to achieve. Meanwhile, users need to find value in their purchases. Moreover, the checkout process is awkward, but we can't change it substantially. How do we then address these goals and needs?

One solution would be to offer a subscription service, charging the full retail price but providing convenient, free shipping. The business thereby increases the number of items purchased through the subscription plan while avoiding conspicuous, off-brand discounts by absorbing the shipping cost instead. The user receives value by saving on shipping costs, as well benefiting from the added convenience of home delivery. Both the business and its users benefit from the subscription service, thereby reducing the number of awkward online checkouts. Everybody wins. What results is a meaningful experience.

Meaningful experiences transform a digital creation into a manicured result for both users and businesses. Users engage. Businesses grow. You spend less time and money achieving these results, as effective UX design and research shows us both what we should create and what we should not. If you try to avoid UX, you may find yourself grasping at unobtainable goals, clawing through unforeseen obstacles, and flailing amid undeniable failures. On the other hand, if you embrace UX, you will likely find the greatest successes are well within your reach.

Key Takeaways

- User experience is the result of using any product or service.
- UX is primarily split between design and research activities.
- The commonality between all UX roles is a focus on users.
- Effective UX design and research saves time and money.

Questions to Ask Yourself

- What are the user goals?
- What are the business objectives?
- Where do user goals converge with business objectives?

You Are Not the User

The Yangtze River stretches nearly 4,000 miles across central and eastern China, feeding from glacial and wetland tributaries as it weaves through the Qinghai–Tibet plateau, passing over the ghostly, submerged towns of the Three Gorges Dam, and emptying into the East China Sea. The river provides a home to many residents, including a remarkable fish called the torafugu (see Figure 2-1). It swims through both the Yangtze's lowland waters and your software projects.

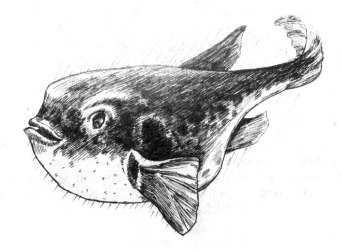

Figure 2-1. Artist's rendering of a torafugu swimming

© Edward Stull 2018
E. Stull, *UX Fundamentals for Non-UX Professionals*,
https://doi.org/10.1007/978-1-4842-3811-0_2

The torafugu, also known as the tiger blowfish, would be unremarkable, save for its two notable features: first, it is delicious and often described as the most flavorful sashimi (similar to sushi); second, it contains a neurotoxic poison called tetrodotoxin that makes cyanide look like salad dressing. A few drops of tetrodotoxin[1] could kill several adults in the most horrific of ways, as it first paralyzes its victims, then it slowly deprives their motionless bodies of necessary oxygen. The poison has no antidote. Luckily, only some parts the fish are poisonous, leaving the rest for the nimble knives of specially licensed sushi chefs and adventurous diners.

Software project teams are like these sushi chefs, capable of creating a masterpiece and of poisoning their customers. With focus and precision, everyone wins. However, without careful attention, poison seeps into the experiences we create.

What is the poison in this scenario? It is the unavoidable bias we introduce into a project: our long-held beliefs, our unfounded opinions, and our hasty generalizations. We carve up a project, and—ever so slowly, ever so unwittingly—biases bleed into the work. We take shortcuts, making decisions based on familiarities and preferences. A familiarity with iPhones may lead us to neglect the needs of Android users. A preference for subtly contrasting colors may lead us to neglect the needs of the color blind. Such biases and countless others permeate our decisions and risk poisoning the experiences we design.

How do we avoid poisoning an experience? We simply recognize that we are not the users of the experiences we design. The phrase "you are not the user" is an axiom in the UX community. At its surface, it stands as an indisputable statement: You are you; you are not someone else. Therefore, we can never truly see an experience through another person's eyes. We can research. We can empathize. But we cannot be users of something we ourselves create.

What Is a User?

If you are new to user experience, you may have noticed that the word "user" tends to be thrown around a lot. You will hear, "user-centered design, user goals, user journeys," and, of course, "user experience." It would seem that the user is highly prioritized within the field of UX. But even within UX, the user is often neglected. Consider the following statement:

> A user is a person having an experience.

---

[1] "Tetrodotoxin : Biotoxin." Centers for Disease Control and Prevention. November 09, 2017. Accessed June 07, 2018. https://www.cdc.gov/niosh/ershdb/emergencyresponsecard_29750019.html.

The statement is so sparse that it sounds whimsical. Yet, this basic idea is at the core of what a user is. Over time, an artifice forms around this definition, complicating its discourse and draining its value. What was once a simple idea branches off into multiple directions, like a river spreading across a delta. You can surround it with marketing flourishes or embellish it with academic phraseology, but the fundamental idea remains: You must have a user to have an experience, and you must have an experience to have a user. The two are inseparable.

You might see yourself as a potential user when creating an experience, such as a website or an app. For instance, your team may create a gourmet cooking app, full of tasty recipes. You think, "I love food, and I know a lot about gourmet cooking." But, even though you may use an app, you are still not the user—you are the *creator* using the app. Even experienced designers sometimes make this mistake. Consider the following hypothetical example:

> Fishes'R'Us wants to create an iPhone app that helps users understand how to cook seafood. You love seafood and cook it often; therefore, you believe you are a user. Sounds logical enough, yes? However, a problem arises in following such logic because, even though you may be a member of the target audience, your mere involvement in the project affects your objectivity. This is best demonstrated by taking the previous example and adding a bit of background information:

> Fishes'R'Us wants to create an iPhone app [*and is paying you to provide a solution*] that helps users [*who may know more or less than you do*] understand how to cook seafood. You love seafood and cook it often [*and you already know how the app works, what it offers, and what it does not*]... Do you still believe you are a user?

You have a vested interest in designing an experience. You want your client to be satisfied, your company to be successful, and your team to be happy. These concerns can affect your objectivity. They often do. Perhaps your client's desire to create an app is misguided, and the app should instead be a website, a Facebook page, or a podcast. Perhaps your company wants to "wow" the client and recommends unnecessary features and functionality. Perhaps you want to be seen as a team player and support your team's cutting-edge ideas. These desires and concerns are understandable, and some may even be admirable.

However, the cruel reality is that users do not share these concerns. Users do not know your client, your team, or you. They do not care about your project as much as you do—if they do at all. They cast their attention toward their own lives, their own needs, and their own desires. Their thoughts are filled with private concerns about their jobs and families, as well as their own projects, ranging from the banality of mowing lawns to the excitement of planning vacations. You may eventually lure users into caring about the experiences you create, but a user's biases and interests will always differ from your own. He or she may learn to love your creation, and he or she may eventually use it every hour of every day, but—right now, at this moment—you are not that person. You are not the user.

Key Takeaways

- Teams unconsciously introduce biases into their work.
- Acknowledging bias helps avoid its effects.
- Teams are creators of an experience, not the users of it.
- Users do not share your concerns about your client, company, or team.

Questions to Ask Yourself

- Am I designing for my users, my client, my team, or myself?
- What vested interests do I have in a project succeeding?
- What information do I know that a user would not?
- Am I expecting too much from users?
- Do I fully understand the needs of users?

You Compete with Everything

Like clockwork, the Summer Olympic Games open with a flurry of fanfare and excitement every four years. Ranging from archery to wrestling, more than 300 individual events are featured.[1] Each day is packed. Athletes, coaches, and fans race from venue to venue, as moments tick away under the burning flame. You could not see every event in person, even if you dedicated all your time to the pursuit. We face the same challenges with many experiences, from watching the Olympic Games to designing software.

Since the beginning of the modern Olympic Games in 1896, more than 100 events have been discontinued, everything from dueling pistols to the standing high jump. A spectator at the 1900 Paris Games watched the grisly spectacle of live pigeon shooting. Off the shores of Southampton, Olympic motor boat racing captivated a handful of lucky onlookers during the 1908 London Games. While some events hold our attention, others fall out of favor over time.

We should not be surprised to learn that when combining the interests of 200 National Olympic Committees,[2] thousands of athletes, and millions of spectators, we are left with a multitude of possible sporting events, not all of which are suitable for the world stage. Pairing down all these choices to a

---

[1]"Sports." International Olympic Committee. January 14, 2018. Accessed June 07, 2018. https://www.olympic.org/sports.

[2]"National Olympic Committees." International Olympic Committee. May 24, 2018. Accessed June 07, 2018. https://www.olympic.org/national-olympic-committees.

© Edward Stull 2018

E. Stull, *UX Fundamentals for Non-UX Professionals*, https://doi.org/10.1007/978-1-4842-3811-0_3

manageable number of competitions becomes an Olympic-level achievement in itself. Likewise, when we create a digital experience, we too need to pare down a multitude of possibilities. Should we add a button here? Should we remove a link there? Each item, should it make our cut, will compete for the user's attention. Cumulatively, these items become one experience in our users' lives. They read a tweet. They visit a website. They launch an app. Users determine which are worthy of their attention and which are not.

One Choice Out of Many

Consider for a moment all the things a person *could* be doing right now. Countless choices exist, from watching TV, to writing the great American novel, to playing with their kids. Along with the things this person could be doing are the things he or she *should* be doing: paying bills, preparing for the following workday, taking out the trash, etc. Lastly, add to these with all the things he or she *would rather* be doing: taking a vacation, eating a good meal, or doing other things best left to the imagination. Now, you and your team want to carve out a bit more time from this person's day. Your creation is worth it. Right?

Getting a person to visit your website or use your app is a minor miracle. If it is a website to share brand information… good luck. An application to connect with loyal customers? Fat chance. Your challenge is not creating a digital experience; your challenge is creating a digital experience that—at the most—will be one tiny part of everything a user could, should, or would rather be doing (see Figure 3-1). Like starry-eyed athletes, project teams strive for the applause of audiences and the notoriety of awards, only to later realize that their true competition was not only their fellow creators, but it was everything that existed at every moment, in every day, of every user's life.

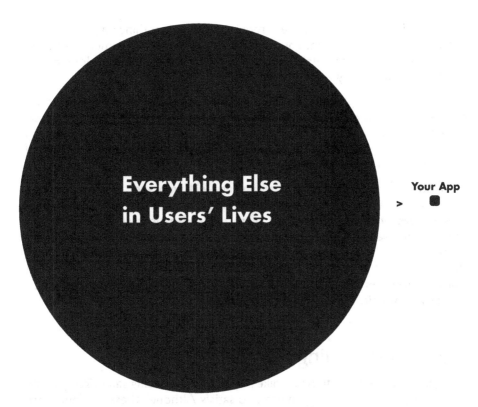

Figure 3-1. Experiences are in constant competition for user attention–both directly and indirectly

Competing with everything is an antithetical statement for many project teams. As a member of a project team we are focused on the act of creation. We know that we compete within market segments: Brand A is better than Brand B, App #1 offers more than App #2, etc. Yet, we are often far less honest with ourselves about how small of an impact our creations make on users' lives. Yes, it would be wonderful if users loved our project as much as we do. But they do not. They do not care if our app sells. They do not care about our industry. They do not care about you or me.

Embrace, Not Accommodate

Users do care about themselves. They seek solutions to their needs. Needs range from locating emergency assistance to satisfying idle curiosity. Necessity. Utility. Entertainment. Companionship. Advice. You name it. Reasons span the vital to the mundane and are only limited by a user's imagination, circumstance, and attention. However, every reason shares a single, common attribute: users would rather embrace a solution than accommodate it.

Try to recall the last time you were required to do something unpleasant, such as fill out a tax return form. You had to find the correct documents. You had to calculate the precise totals. You had to file by a specific date. None of this was likely done joyfully—you accommodated.

Now, try to recall a pleasant situation, such eating a bowl of salted caramel ice cream. You may have considered consuming on a bowl of broccoli, noshing on a plate of pinto beans, or devouring a saucer of semolina, but ultimately you chose what you wanted to experience. You did not accommodate a bowl of ice cream—you embraced it.

How do we create experiences that users will embrace? We already have the answer: users embrace what they willingly choose above all else—what they believe will best meet their own needs. If a user seeks information, she will choose what she believes is the most informative. If a user seeks entertainment, he will choose what he believes is the most entertaining. If they cannot find what they need, they may accommodate a solution, but that solution will never take home a gold medal. It only temporarily satisfies until a stronger competitor enters the marketplace.

The Never-Ending Game

Designing experiences often feels like playing a never-ending game, fraught with high hurdles to jump and selfish users to satisfy. Although these challenges are daunting, you are no more disadvantaged than anyone else. Each experience competes with all others. User experience is a broad but equal playing field, daring all players to strive for greater knowledge, and inspiring all audiences to seek out better experiences.

Key Takeaways

- Users could, should, and would rather be doing a multitude of activities.

- Users should embrace your solution, not accommodate it.

- Users embrace experiences that they willingly choose above all others.

Questions to Ask Yourself

- Do I clearly understand the problem I wish to solve?

- How are users currently handling the problem I wish to solve?

- What products and services are similar to what I am creating?

- What else—both commercial and personal—is competing for my users' time, money, and attention?

- Do I recognize the real impact I am making in users' lives?

- Do users embrace or accommodate what I create?

- How do I create an experience that users will embrace?

The User Is on a Journey

Each year, Pauillac, a village nestled within the Médoc region of France, hosts a marathon. The Marathon du Médoc weaves through 44 kilometers of bucolic Bordeaux countryside. Points along its route include the iconic vineyards of Château Lafite Rothschild, Lynch-Bages, and many others. Green, combed hillsides of grape vines meet revival architecture capped in spires and surrounded by manicured gardens. Race day begins with a fashion show and ends with a fireworks display. Festivities throughout the morning and afternoon entertain onlookers, but each pales in comparison to the main event. If you run this marathon, you will have an unforgettable experience. If you *study* this marathon, you will learn a lot about user experience design.

The Marathon du Médoc is unconventional. For starters, the marathon's atmosphere is relaxed. Competitors are given six and a half hours to finish, which gives them about 15 minutes to complete each mile of the course. Duration, not distance, measures the so-called, "longest marathon in the world."[1] Some runners cheat and start halfway. Others hide bicycles along the path. A few dress up as comic book characters, nuns, or the Village People. And nearly everyone is a little bit drunk.

---

[1]"Marathon Des Châteaux Du Médoc." Marathon Du Médoc. Accessed June 07, 2018. `http://www.marathondumedoc.com/en/`.

© Edward Stull 2018
E. Stull, *UX Fundamentals for Non-UX Professionals*,
https://doi.org/10.1007/978-1-4842-3811-0_4

Approximately 8,500 runners begin their journey with a glass of wine. The aid stations all supply additional glasses of red and white wine, as well as oysters and steak. Completing any marathon is a remarkable achievement—even more so with a belly full of Cabernet Sauvignon and entrecôte. Perhaps unsurprisingly, several hundred runners do not finish the race. The runners were certainly motivated, possibly inebriated, able to cheat without consequence, and presumably don't get lost. So why do some quit? It is hard to say. (A long run fueled by alcohol and shellfish stands as one possible reason.) A runner may encounter a sudden injury or slowly run out of steam under the baking hot, French, summer sun.

A person quits an experience in the same way that a marathoner quits a race: either suddenly or over the course of time. People get distracted. Life gets in the way. Stuff happens. And, although some people will quit, others will succeed. Their experiences unfurl across a figurative landscape, crisscrossed by pathways and populated by a mix of happenstance and design.

Many experiences, such as a marathon, may first appear as a linear path—start to finish. But, upon further discovery, they reveal themselves to be complex journeys, beginning in several places and ending in many possible outcomes.

A helpful way to think about a user journey is to imagine a marathon full of drunk runners. The runners wish to reach the finish line, but they are susceptible to exhaustion and easily distracted. Their senses are dulled, and they are unaware of how to reach their goal.

The road they travel on contains several intersections. Upon encountering one of these crossroads, runners must either continue on their current course or make a turn (see Figure 4-1). Without guidance, they stumble around looking for clues about how to proceed. Such moments may determine the success or failure of a runner's entire race, keeping him or her on-track or steering him or her off-course.

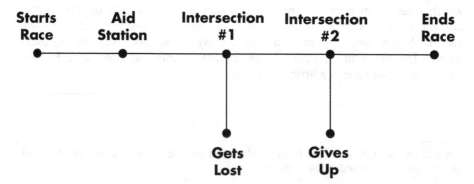

Figure 4-1. Users may continue or abandon a journey at any moment

Aid stations along the way provide brief respite for the runners, where they receive help, swig wine, and slurp oysters. However, if the runners stay too long, their bellies fill and their legs grow tired. Too much aid can be a bad thing when a person pursues a goal.

Now, instead of a drunk person running down a road in France, imagine a person buying a plane ticket online. This buyer may be unaware of how to reach her goal. She may become distracted or give up out of exhaustion. The journey she takes contains several intersections, as well (see Figure 4-2). She visits a website, enters a credit card number, and receives an email confirmation. At any one of these crossroads, she could choose another path. For example, visiting a website may lead her to complete her purchase over the phone. Entering a credit card number could generate a confusing error and cause her to shop elsewhere. Her email confirmation may entice her to join the airline's loyalty program. Each of these events serves as a potential off-ramp from one experience and an on-ramp to another.

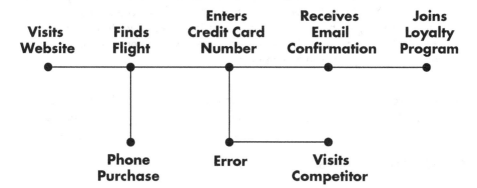

Figure 4-2. A user's journey may encounter detours and distractions

We can offer help to users during such a journey (see Figure 4-3). However, like an all-you-can-eat aid station, too much assistance can slow down a user's pursuit of his or her goals. Repeated alerts dull their senses. Long explanations nauseate rather than alleviate. We must not place too many treats on the table, lest we lose users to the comfortable apathy of effortless abandonment. The easiest choice a user makes is doing nothing at all.

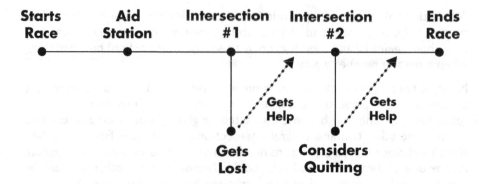

Figure 4-3. Ideally, we offer just enough help to keep users moving toward their goals

So, how do we design experiences based on a user's journey? We cannot force a user to do anything, but we can pave the way to preferred outcomes. We do this by removing obstacles, planning detours, and offering guidance when needed. Yet, knowing where to focus our efforts often proves to be the biggest challenge when designing an experience. We can determine these locations by studying three things: where a user was, where the user is, and where the user is going (see Figure 4-4).

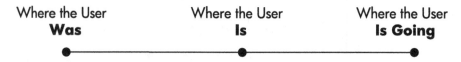

Figure 4-4. The three states of a user's journey

Where the User Was...

The user begins our journey at the first intersection along the road, where her path and ours meet. After all, she comes from somewhere else. She has walked other roads before ours. When we meet the user at this crossroad, she either decides to join us or ignore us. More often than not, she fails to even take notice—many distractions compete for her attention.

We want to know where the user was before we met her. We want to know her *context*. The user's context is arguably the most important part of a user's journey, for it often determines which path she will take next. If the user comes from a context applicable to the path we constructed, she may join us; if not, she will likely take another route. For instance, buying a plane ticket online would be applicable in the context of planning a vacation, whereas it rarely would be in the context of planning a meal.

Where the User Is…

With a bit of coaxing, the user elects to travel down our road. Her journey with us will continue or end at the next intersection. For instance, she will reach a crossroad where she will debate buying from us.

We can guide the user to a preferred path if we know where she is within the journey. If she is ready to learn, we should teach. If she is ready to buy, we should sell. But, if she is not ready for an experience, anything we say or do about it will be misplaced.

Too often, a misplaced experience leads users astray when a request comes too early or an incentive comes too late.

For example, some apps request permissions (see Figure 4-5) as soon as they launch, such as a dialog that reads, "Allow ACME to access your location." The request asks users to choose among "Only while using the app", "Always Allow", or "Don't Allow." Although such behavior provides flexibility for the following experiences, it only does so after the user has accepted. Acceptance requires trust, and a new app upon its initial launch is unlikely to have earned it yet. If the user denies the request, the following experiences may be crippled. Like a tall hurdle placed in the middle of a road, some users skip the request to jump and simply walk around it. The request came too early.

Figure 4-5. An iOS11 dialog requesting permission to access the user's location

I once worked with a major online retailer that offered free shipping on their checkout page. Free shipping is a proven conversion incentive[2], but users view checkout pages only after they decide to buy. Like an aid station that is just out of sight over the next hill, most users never saw the incentive. The incentive came too late.

We can predict these intersections by researching how other users behave. Each user is unique, but groups of users tend to follow similar paths along a journey, thereby allowing us to anticipate where a user may confront an obstacle, make a detour, or veer off course into the vast wilderness of countless possibilities.

Where the User Is Going...

If our research is correct (and with a bit of luck), we can anticipate the intersections, points of interest, and dead ends along a user's journey. We design accordingly.

Each design decision becomes a result of the user's goals. If we know that users first search for a product, we should direct our awareness-building efforts toward search engine marketing and optimization. If we believe users make purchase decisions only after an app's download, we should focus on post-installation conversion. If we understand that users abandon their accounts within 90 days, we should foster retention within the first few months of use.

Despite their many similarities, a marathon full of drunk runners and the journey of users do differ in at least one key way: running a marathon is a solo act, driven by the skill and passion of an individual. In contrast, a user journey is a partnership between a user and a designer, driven by the designer's empathy for the user and an understanding of the user's goals.

The more you understand a user's goals, the greater the chances are that you will reach yours as well. A user journey is merely the road map to achieve them. Place yourself in the user's shoes and design the paths that he or she will travel. It is the only way to win. So, open a bottle of wine, crack open a few oysters, and sizzle up some steak—you have a marathon to run.

---

[2]Roggio, Armando. "The (Many) Benefits of Offering Free Shipping." Practical Ecommerce. May 31, 2018. Accessed June 07, 2018. https://www.practicalecommerce.com/The-Many-Benefits-of-Offering-Free-Shipping.

Key Takeaways

- The easiest choice a user makes is doing nothing at all.

- Context affects a user's behavior.

- Effective UX anticipates events a user may encounter.

- Understanding where users are within a journey allows designers to guide users to beneficial outcomes.

Questions to Ask Yourself

- What obstacles can I remove from an experience?

- What assistance may I provide to users?

- Am I assisting users or burdening them with too much help?

- If a user abandons an experience, where does she go to next?

- How can I preserve user safety, security, and dignity throughout an experience?

Keep It Simple

German engineers designed a terrifying tank in the closing years of World War II. They called it the Maus[1]—or "Mouse" in English. The ironically named tank dwarfed its Allied competition in both size and ferocity (see Figure 5-1). However, like most complex creations, it could win a battle but not a war.

Figure 5-1. Artist's rendering of the Panzerkampfwagen VIII Maus super-heavy tank

---

[1]Heinz Guderian, *Panzer Leader* (Heidelberg: Kurt Vowinckel Verlag, 1950).

© Edward Stull 2018
E. Stull, *UX Fundamentals for Non-UX Professionals*,
https://doi.org/10.1007/978-1-4842-3811-0_5

The Maus weighed over 200 tons,[2] as much as a blue whale. The tank cracked street pavement while moving over it and sank into muddy ground when standing still. Bridges crumbled under its tracks. The tank's weight afforded its crew safety, wrapping its occupants within a nearly impenetrable, eight-inch thick, welded and cast steel shell. Atop its chassis sat a massive, 128 mm gun,[3] adapted from naval artillery. Half as long as a telephone pole, the gun's barrel fired a huge, 62-pound projectile at nearly three times the speed of sound. Combined with the tank's armor, the Maus could defend against and destroy anything on a battlefield.

In comparison, the American Sherman tank looked downright tiny, being 15 times lighter and having half the caliber (see Figure 5-2). It proved to be reliable and quick, but the tank was lightly armored and under-gunned. Adding to these deficiencies, the Sherman's engine used highly flammable, aircraft-grade fuel. Once hit, the Sherman tanks frequently erupted into flames, earning them the unfortunate nickname of "Tommy Cookers".

Figure 5-2. An American Sherman tank landing on a Sicilian beach in 1943[4]

---

[2]Intelligence Bulletin. Publication. Washington, D.C.: Military Intelligence Service, 1946. http://www.lonesentry.com/articles/maus/index.html.
[3]Terry Gander and Peter Chamberlain,. *Weapons of the Third Reich: An Encyclopedic Survey of All Small Arms, Artillery and Special Weapons of the German Land Forces 1939–1945* (New York: Doubleday, 1979).
[4]Photo by Signal Corps, "Sicily Invasion," July 10, 1943.

If given the choice, would you choose a massively armored, nearly impenetrable tank that dominated its opponents, or would you choose a lightly armored, under-gunned tank that burned its crew?

The Maus is the better option until you consider its only shortcoming: the Germans never finished building one. Plagued with mechanical issues, poor crew training, and bombarded manufacturing facilities, the terrifying roar of the Maus barely even made a squeak.

Ultimately, the Maus tanks were too complex to assemble and too complicated to maintain. Two hundred tons of complex mechanics require vast amounts of resources and time. When Russian soldiers captured the Maus' proving grounds, they found only two partially built tanks. Luckily for the Allies, as well as history, the aspirations of German designers were never fully realized.

In contrast, American factories mass-produced the Sherman, churning out nearly 50,000 tanks. So abundant were the tanks that stories[5] emerged of eight or more Shermans swarming a single German opponent, surrounding it from all sides, creating a ring of cannon fire. Placed end-to-end, Shermans could encircle the entire capital city of Berlin. Mighty or not, the Maus could not fight math.

Head-to-head, a battle between a single Maus and a single Sherman would certainly favor the Germans. The American tank's simplicity was both its weakness and its strength, because wars are not won in single tank duels—they are won by design.

What does this decades-old example teach us about design? Complex designs are hard to build and even harder to maintain, be it a World War II tank or tomorrow's mobile app. No design exists in a vacuum. Each one is affected by all the others, and each one is only a small part of a greater system.

Direct comparisons of competing products lead us into a never-ending arms race of features. Like a pair of dueling tanks, a single comparison may favor a complex solution over a simple one, but we must consider the entirety of a user's experience to understand which product will ultimately win. Who is our user? What is she attempting to do? When does she do it? How is she currently coping without our product? Why is our solution better than the second-best alternative?

The wreckage of feature-rich products litters software history—Microsoft Bob, Google Lively, iTunes Ping to name but a few.

---

[5]Armor in Battle. Publication. Fort Knox: U.S. Army Armor School, 1986. http://www.benning.army.mil/Library/content/mcoelibrarieseresources/ebooks/Armor%20in%20Battle_FKSM17-3-2.pdf.

When we continually add to our creations, we weigh them down with complexity. We create complexity through the act of creation itself. It comes in the form of ideas, budgets, schedules, briefs, proposals, presentations, screens, gestures, web services, repositories, databases, scripts, classes, structures, and bug reports.

We roll out our software and hope it survives among the thousands of other experiences competing for our users' attention. Yet, complexity obliges us to focus our efforts on the construction and maintenance of the software itself, instead of the experiences it creates. We lose sight of our objectives as we pursue the grand, the expansive, and the robust. Features break. Support fails. Team members leave. Such flare-ups and conflicts divert us from our one true goal: we wish to fulfill a user's need in the simplest possible way.

Let's look at three methods to defeat complexity.

Absence

In the story of Adam and Eve, a serpent tempted Eve to take a bite from the forbidden fruit. Eve could have been tempted by any number of distractions, ranging from harp lessons to finding sunblock. Yet, a talking snake grabbed her attention. The snake offered Eve knowledge. Eve accepted, and she and Adam were kicked out of Eden.

From a user experience perspective, we cannot blame Eve. She is our user. Users always crave knowledge, especially when they are tempted by something new and exciting. Think of the countless times you've ventured online to buy a gift, only to be derailed by a BuzzFeed article. We could easily blame the snake, but he is only partially to blame. The snake merely directs Eve toward the problem. He is not the problem itself. The problem is the forbidden fruit. Remove the fruit and the problem is solved. Adam and Eve lounge around for eternity, occasionally striking a pose for a Michelangelo fresco.

Like the removal of the forbidden fruit from the Garden of Eden, we, too, can remove a distraction from our creation before it becomes a problem. People will not get themselves in trouble if you take away the opportunity to do so. Applications already do a great deal of work on behalf of users. Users do not need to pick the cell towers through which their calls are routed. They do not need to tell a website to encrypt their passwords. They do not need to translate video game moves into machine code. Why should users need to press a button? Why click a link? Why check a box? Why should users be required to do anything at all? When it comes to software features, absence is underrated.

Reduction

Wilderness firefighters chop down trees and clear brush ahead of an approaching fire. They blaze a perimeter, called a control line, to remove combustible materials (see Figure 5-3). Once the fire reaches the control line, it runs out of fuel. The absence of fuel suppresses a fire. Everyone goes home.

Figure 5-3. Firefighters burning a control line[6]

An application with unnecessary features becomes dangerous over time. It withers and takes a spark like a field of dry grass. When we seek limitation, we remove its fuel.

For example, you may learn that users tend to abandon a website's form halfway through, only filling out five of its ten fields. If you were to remove the one field, you may improve the form's performance slightly. But if you were to remove five fields, then all the users would complete the form. If an experience stops being successful halfway through it, remove the last half. The quickest way to achieve success is to stop when you find it.

---

[6]Stevepb. Grass Fire. Digital image. Pixabay. June 16, 2015. Accessed June 6, 2018. https://pixabay.com/en/grass-fire-firefighter-smoke-807388/.

More often than not, reduction simplifies what remains.

- Need to emphasize a message? Shorten it.

- Want to increase the number of form submissions? Decrease the number of fields.

- Want visitors to email you? Remove your phone number.

- Want something to look less expensive? Delete the $ sign before the price.

Addition

The 19th-century Russian writer and playwright Anton Chekhov once instructed, "If you say in the first chapter that there is a rifle hanging on the wall, in the second or third chapter it absolutely must go off. If it's not going to be fired, it shouldn't be hanging there." It would serve us well to follow his advice when designing applications.

The key difference between a good addition and a bad distraction is what is being added: answers, not questions. Consider how the following additions may improve an experience:

- Asking users a yes-or-no question? Preselect yes.

- Need users to choose a date? Set the default date to today.

- Want users to share something? Supply the message.

Can your application do something on behalf of the user? If so, do it.

In the correct context, added complexity can improve user experience. It is commonsense. A song composed of a single musical note would grow tiresome. Likewise, unseasoned food is simple, but often unpalatable. Consider the play mechanics of video games. Here, designers add complexity, intentionally obstructing players as they pursue their goals. In Activision's *Call of Duty*, a player faces thousands of challenges, ranging from avoiding UAVs to decapitating zombies. Players get shot, crushed, and incinerated. And they welcome it. We can view these complexities as beneficial because the challenges are enjoyable.

So, we can understand our true battle is one where we manage complexity through absence, reduction, and sometimes even addition. Managed complexity has the power to inform and entertain. Unmanaged complexity confuses and distracts. It's the tank that cannot be built. It's the forbidden fruit. It's the field of dry grass. It's the gun hanging on wall, waiting to be fired at a problem that does not exist.

Key Takeaways

- Complex designs are difficult to build and maintain.

- Other products and services compete for your users' time and attentions.

- Unnecessary features distract users away from necessary features.

- Provide users a default solution and allow users to edit it as needed.

- Good UX is more than a summation of features.

Questions to Ask Yourself

- What value does an experience provide to a user?

- What features can I safely remove from an experience?

- What tasks can I do on the behalf of users?

Users Collect Experiences

The Prince of All Cosmos shines as the star of Namco's 2004 hit video game *Katamari Damacy*.[1] He is an ambitious little fellow. At only two inches tall, the tiny prince furiously runs in place on top of a huge ball. The ball moves across the Earth's surface like a snowball, collecting everything it touches. Growing ever larger, the spherical cluster of strange objects picks up ants, thumbtacks, fence posts, sumo wrestlers, cows, bulldozers, cruise ships, clouds, monuments, and buildings (see Figure 6-1). The prince's appetite knows no bounds. He gathers everything he can find to complete his goal. As users, we do the same thing every day.

---

[1]Katamari Damacy. Computer software. San Jose, CA: Namco, 2003.

© Edward Stull 2018
E. Stull, *UX Fundamentals for Non-UX Professionals*,
https://doi.org/10.1007/978-1-4842-3811-0_6

Figure 6-1. Artist's rendering of Katamari Damacy ball

We share much with the prince, because we, too, collect what we experience. We pick up bits and pieces. Everything you see, touch, smell, and taste becomes a part of you, reshaping your expectations as you travel across a landscape full of possibilities. For example, our expectation of an elevator button is affected by our experiences with microwave keypads, game console controllers, mobile phones, and any other buttons we've encountered.

We anticipate that financial websites will compute, video games will entertain, and weather apps will forecast. Yet, these are base-level expectations that are continually evolving. Each experience informs the next. And, like a snowball, our experiences grow into an interlocking network of adjacencies, assemblies, and conglomerations. What we experience today is the indirect result of every other prior experience, be it playing a video game or designing software. We soon realize our best ideas are often merely combinations of the past and present.

Even the imaginative gameplay of *Katamari Damacy* is an amalgamation of traditional and contemporary activities. The game behaves like tamakorogashi,[2] a sport where a large, rolling ball is steered by Japanese schoolchildren during an outdoor race. You can find inspirations within the game from *Pac-Man* to *Super Mario Bros.* to *Final Fantasy*.

How do we combine ideas to better serve our users? First, we need to understand what the users have collected thus far: we need to understand their context.

Context

Every experience consists of an event, a time, and a place. An event is what happens. For example, a ball rolls. Time is when an event happens. For example, a ball *starts* rolling. A place is wherever the event happens. For example, a ball starts rolling *down a hill*. If we were to design a new experience involving this ball, we would want to know all the past experiences that have led a user to our chosen event, time, and place. In short, we want to understand the user's context.

Context affects a user's ability to appreciate a designed experience. A mobile app that works in an office setting may not work when riding a bicycle. Checking into a location makes perfect sense to Foursquare and Facebook users, although it may bewilder other audiences. A Cancel button on a Mac dialog appears before an OK button, but the opposite is true on a PC.

Consider the context of a user placing an item in a shopping cart. Does she expect to see a sign-in or register before checking out? If so, why? Perhaps she thinks all credible websites have them. Is not having a PayPal option a problem? Why? Maybe she doesn't own a credit card. Does she look for a lock icon to determine if a page is secure? Why? Many people may judge security this way. Only after understanding a user's context can you design her experience.

An experience will either stick or bounce off, depending on what a user has already gathered. We need to provide something useful, meeting our users at exactly the right time and place. The world is big, and we are only a small part of it. At best, we can fulfill an individual need, at a specific time, in a particular place. Doing so requires context. We must keep our eye on the ball.

---

[2]Sheffield, Brandon. "Postcard from GDC 2005: Rolling the Dice: The Risks and Rewards of Developing Katamari Damacy." Gamasutra Article. March 11, 2005. Accessed June 07, 2018. https://www.gamasutra.com/view/feature/130660/postcard_from_gdc_2005_rolling_.php.

Key Takeaways

- Past experiences shape future experiences.
- Every experience consists of an event, a time, and a place.
- A user's context represents a culmination of all his or her past experiences.

Questions to Ask Yourself

- Where was the user before?
- Where is the user now?
- Where will the user go next?

Speak the User's Language

In 1799, a young French lieutenant named Pierre-François-Xavier Bouchard made one of the greatest discoveries of all time, only to lose it two years later to the British.[1] His discovery was neither golden nor bejeweled. However, it has mesmerized kings and scholars, generals and diplomats, readers and writers, for centuries. It is also a fine example of user experience design.

Bouchard's discovery resided in the Egyptian port city known today as Rashid, located nearly 2,000 miles from his birthplace in Orgelet, France. Rashid had long been a desirable center of trade and commerce, for it lay on the Nile River's banks and was cooled by the gentle winds of the Mediterranean Sea.

At the city's edge stood Fort Julien. Its crumbling walls included a patchwork of earlier fortifications and repairs, one of which Bouchard uncovered while excavating a wall's foundation. The discovery was a *stele*: an inscribed, ancient volcanic stone slab.[2] At nearly four feet high, three feet wide, and a foot thick, the slab provided a stable—if not somewhat underappreciated—support for the wall above. It had been placed within the stone wall and hidden by sand, dirt, and time. Two thousand years passed between its inscription and excavation, all the while it held the secret to understanding a long-forgotten world.

---

[1]"Everything You Ever Wanted to Know about the Rosetta Stone." The British Museum Blog. August 02, 2017. Accessed June 09, 2018. https://blog.britishmuseum.org/everything-you-ever-wanted-to-know-about-the-rosetta-stone/.

[2]Saunders, Nicholas J. *Alexander's Tomb: The Two Thousand Year Obsession to Find the Lost Conqueror*. New York: Basic Books, 2006.

E. Stull, *UX Fundamentals for Non-UX Professionals*,
https://doi.org/10.1007/978-1-4842-3811-0_7

Carved in 196 BCE,[3] the stele detailed a list of the good deeds that the king of Egypt, Ptolemy V, had performed for temples and people in the region. It likely stood upright in a temple or public area, and may have been a part of a much bigger stone, as the stele's message was cut off by chips and fractures. What we can read describes how the king increased gifts and reduced taxes. It spoke of how the king's armies vanquished their enemies. It told of how the gods granted the king "strength, victory, success, prosperity, health, and other favors." Rulers have always enjoyed telling people of such things. In essence, this stele served as a form of advertising: a chiseled billboard meant to sway the opinions of passersby.

Its creators inscribed three languages onto the stele, forming rows of bright white markings offset against the stele's dark gray surface. Each of the three languages targeted a different audience: hieroglyphs spoke to the priests, demotic to the common people, and Greek to the ruling class. The king needed to speak to each group in their preferred way, lest his message go unreceived. (Even kings have requirements they must meet.) Although the languages differed, the messages of all three were the same: I understand your needs; here is what I've done to fulfill them.

Today's linguists can decipher Egyptian hieroglyphs, but the writing system had disappeared into obscurity by the time Bouchard viewed its strange pictographic shapes and symbols of delicate birds, outstretched snakes, and solemn eyes. Archeologists had been unearthing artifacts covered in these pictographs for decades before Bouchard, but this stele was unique. Until its discovery, this strange language had not been displayed alongside a Greek translation. Scholars already understood Greek letters; but they did not yet understand Egyptian hieroglyphs. In the years that followed, they used the known language to decipher the unknown one, thereby unlocking the mystery of the stele we now call the Rosetta Stone (see Figure 7-1).

---

[3]"The Rosetta Stone." British Museum. Accessed June 07, 2018. http://www.britishmuseum.org/research/collection_online/collection_object_details.aspx?objectId=117631&partId=1.

Figure 7-1. The Rosetta Stone[4]

As makers of digital work, we create new languages that users—the people who use our creations—must decipher. The languages we create may not be as ornate as hieroglyphs, but they are languages nonetheless. Marketing, graphic design, and technology (to name but a few) are, to most audiences, as cryptic as an ancient language. Marketing speaks of equity and segments. Graphic design speaks of balance and harmonies. Technology speaks of stability and performances. Depending on your career and interests, one of these languages may be more familiar than the others. Perhaps you are even fluent in all of them, but I would hazard to guess that the users of what you create are not. People will still need to understand what you build. Like an ancient king, we, too, have requirements we must meet.

---

[4]Djorenstein. Rosetta Stone. Digital image. Pixabay. January 6, 2017. Accessed June 6, 2018. https://pixabay.com/en/rosetta-stone-heiroglyphics-language-1958394/.

What is the known language that allows a person to decipher all others? With the Rosetta Stone, scholars used their knowledge of ancient Greek to decipher the hieroglyphic writing. However, users must tap into a larger language. It is the one they have built over their lifetimes—their own user experiences. Every website they've used, every app they've downloaded, every device they've held, every video they've watched, every item they've bought, every community they've joined, every culture they've embraced, every lesson they've learned, every success, and every failure is a part of their own user experience. In turn, they use these experiences to decipher the new ones they encounter.

User experience design serves a similar purpose to the Rosetta Stone; it transforms the unknown into the known, translating the many cryptic languages of business into a single, cohesive experience for the user. It makes design immersive, marketing engaging, and technology invisible. Like a message written in a native tongue, a designed experience is understood by users as easily as if it were composed by the users themselves. Such experiences can assume many forms, ranging from apps to websites, but what remains each time is the same message to the user: I understand your needs; here is what I've done to fulfill them.

Key Takeaways

- People use their past experiences to decipher new information.

- Users come from diverse backgrounds and may not understand business, marketing, and technology.

- UX translates business, marketing, and technology solutions into meaningful experiences for users.

- Effective UX fulfills users' needs.

Questions to Ask Yourself

- How can I make an experience more applicable to a user?

- Does an experience require specialized knowledge to use effectively?

- What prior experiences of users can I leverage to make an experience intuitive and familiar?

- If users designed the experience themselves, how would it differ from my solution?

Favor the Familiar

Michigan J. Frog was unlike any other frog. He sang. He danced. He was destined for stardom. In Warner Bros.' 1955 cartoon, *One Froggy Evening*, a construction worker freed Michigan from a time capsule buried within a recently demolished building's cornerstone. Upon reaching the open air, the frog stood and sang, "Hello my baby, hello my honey. Hello, my ragtime gal. Send me a kiss by wire. Baby, my heart's on fire." The construction worker gasped in amazement as Michigan, wearing his trademark top hat and matching cane, pranced across the lid of the time capsule, which moments before had been the frog's boxy prison.

The construction worker fantasized about the riches he could earn by having Michigan perform in front of adoring crowds. But, as he would soon learn, the frog refused to perform for anyone other than his rescuer. Every time the construction worker would show off Michigan, the frog would simply ribbit and croak. No singing. No dancing. No adorning crowds.

How often have you felt the same? "Users are going to love this idea," you say. "They have never experienced anything like it before." You eagerly build out your product, feverishly crafting every exquisite detail. Thoughts of grandeur race through your head—your product will be celebrated. Perhaps even taught in schools. You polish. You finish. You release. And… nobody uses it. Ribbit.

Why does this happen? We create a new product, desiring to make something different and innovative. But we must ask ourselves a critical question: do users share this desire?

© Edward Stull 2018
E. Stull, *UX Fundamentals for Non-UX Professionals*,
https://doi.org/10.1007/978-1-4842-3811-0_8

The Curse

Our familiarity with products can lead us astray. We have a cognitive bias, where we sometimes believe that everyone knows what we know. This "curse of knowledge" was first described by Colin Camerer, George Loewenstein, and Martin Weber.[1] Although their research pertained to economics, the curse of knowledge affects everything from classrooms to mobile apps.

Place yourself in the shoes of a novice user. Pick any topic unfamiliar to you—for instance, aerospace engineering, rail transport, or constitutional law. If you visited a website about the topic, what would you expect to see? What makes it new or different?

Chances are, when dealing with unfamiliar topics, people neither recognize what is typical nor do they desire something different. After all, new experiences are inherently different. How can users want the unknown? When designing experiences, our expertise can blind us from the needs of users, as they may have little to no knowledge of what we have created. Our desire to innovate outpaces a user's need to merely catch up.

Users adopt technologies according to a bell curve. First expressed by Everett Rogers in 1962,[2] a small fraction of users—about 2.5%—adopts new technologies initially (see Figure 8-1). They are innovators. Over time, these innovators lead to early adopters, which grow to early majorities (34%). To reach the early majority of users, we must first cross a chasm.

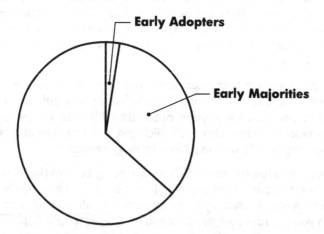

Figure 8-1. The small fraction of early adopters leads to early majorities

[1]Colin Camerer, George Loewenstein, and Martin Weber, "The Curse of Knowledge in Economic Settings: An Experimental Analysis," Journal of Political Economy 97, no. 5 (Oct., 1989): 1232-1254. https://doi.org/10.1086/261651.
[2]"Diffusion of Innovation Theory." The Theory of Planned Behavior. Accessed June 07, 2018. http://sphweb.bumc.bu.edu/otlt/MPH-Modules/SB/BehavioralChangeTheories/BehavioralChangeTheories4.html.

In *Crossing the Chasm* by Geoffrey A. Moore,[3] the author describes the challenges of adopting high-tech products. Whereas early adopters may accept risks, early majorities are far more pragmatic. They buy from market leaders. They want proven reliability. They resist all else. Acceptance may take years, sometimes only occurring after several failed attempts.

Dropbox serves 500 million[4] users today; however, Palm introduced file syncing in 1997. Spotify currently values at $30 billion;[5] yet, Xerox PARC experimented with music streaming in the mid-1990s. Apple's iPad Pro is emblematic of mobile computing; nonetheless, GO Corporation pioneered a pen-based tablet in 1987.

From Palm Pilots to primordial iPads, people often resist new experiences. But, over time, these products and services grow increasingly commonplace. New becomes familiar.

Affordance

Familiarity takes many forms. In *The Ecological Approach to Visual Perception*, James Gibson describes his theory of how creatures see their environments. For example, the environment could be a swamp, and the creature, a frog. Weighing less than an ounce, a common frog will happily sit on a lily pad. The pad holds the frog's weight, whereas the water surrounding it would not. We can describe the lily pad as "sit-able." It affords the ability to be sat upon by a frog. Gibson calls this ability an affordance.

In 1998, Don Norman wrote about perceived affordances in his groundbreaking book *The Psychology of Everyday Things* (later renamed *The Design of Everyday Things*). He described how design affects our perceptions and interactions with objects. For example, a chair is "sit-able" based on its design: a chair mirrors the proportions of a human body, including the shape of its seat, the width of its arms, and the height of its legs.

---

[3]Moore, Geoffrey A. *Crossing the Chasm*. Chichester: Capstone, 1998.
[4]Darrow, Barb. "Dropbox Touts 500 Million Users But How Many Pay?" Fortune. March 7, 2016. Accessed June 07, 2018. http://fortune.com/2016/03/07/dropbox-half-a-billion-users/.
[5]Roof, Katie. "Spotify Opens at $165.90, Valuing Company at Almost $30 Billion." TechCrunch. April 05, 2018. Accessed June 07, 2018. https://techcrunch.com/2018/04/03/spotify-opens-at-165-90-valuing-company-at-30-billion/.

In the years since, affordances and signifiers (i.e., cues) have become the primary means through which digital experiences are understood. We view an interface (see Figure 8-2) and wonder what is "click-able," what is "scroll-able," and what is "swipe-able," as we wade through a morass of toolbars, sliders, check boxes, tabs, accordions, tooltips, dropdown lists, breadcrumbs, carousels, toggles, radio buttons, text inputs, and links. An interface that is easy to use is often one that is easy to recognize. Familiarity shapes its contours, as prior experiences inform new ones.

Figure 8-2. Several interface guidelines, including Apple Human Interface Guidelines[6], Google Material Design[7], and Microsoft Windows Design[8]

Familiarity creates both restrictions and opportunities. It imposes a boundary but grants us a common reference point at which to begin an experience. It is the comfortable known, instilling us with the confidence to pursue the unknown. Like frogs hopping from lily pad to lily pad, we must trust the landing before we leap.

---

[6]Apple Inc. "Human Interface Guidelines." Purchase and Activation - Support - Apple Developer. Accessed June 15, 2018. `https://developer.apple.com/design/human-interface-guidelines/`.
[7]"Design." Material Design. Accessed June 15, 2018. `https://material.io/design/`.
[8]"Design Applications for the Windows Desktop." Meet the Evangelists. Accessed June 15, 2018. `https://developer.microsoft.com/en-us/windows/desktop/design`.

Key Takeaways

- Our expertise creates a "curse of knowledge," which may blind us to the needs of users.

- New experiences become familiar over time.

- Early adopters accept risks that later-adopting users may not.

- People often resist unfamiliar experiences.

- Signifiers indicate the affordances within an interface.

Questions to Ask Yourself

- Are my users early adopters or latecomers?

- Does an experience require users to have prior knowledge?

- How am I addressing the needs of new users?

- How am I addressing the needs of existing users?

- What cues do I provide to users to help them fulfill their goals?

- Will users understand what is actionable within an experience?

- How can I make an experience more familiar to users?

Stability, Reliability, and Security

On September 27, 1997, the USS Yorktown, a Ticonderoga-class missile cruiser, 567 feet long and 34 feet wide, drifted to a silent standstill off the coast of Virginia[1] (see Figure 9-1). Moments before, a crew member had hit zero on his keyboard, inadvertently triggering a software bug and a subsequent cascade of system failures, including the ship's propulsion. The ship sat dead in the water under the stars for over two hours—a billion dollars' worth of American might defeated by a single keystroke.

---

[1]SlabodkinJul, Gregory. "Software Glitches Leave Navy Smart Ship Dead in the Water." GCN. July 13, 1998. Accessed June 07, 2018. https://gcn.com/Articles/1998/07/13/Software-glitches-leave-Navy-Smart-Ship-dead-in-the-water.aspx.

© Edward Stull 2018
E. Stull, *UX Fundamentals for Non-UX Professionals*,
https://doi.org/10.1007/978-1-4842-3811-0_9

Figure 9-1. USS Yorktown[2]

The software bug was caused by an avoidable division-by-zero error. Take any number and divide it by zero. You get an undefined value. Mathematicians have dealt with such problems for hundreds of years, going back to at least 1734[3]. However, the Yorktown's problem went unnoticed until the error rippled throughout the ship's control center network, destabilizing every connected machine like a seismic sea wave. Systems went offline. Engineers scrambled.

We notice stability only in its absence. Stability is invisible. It is the ship that sails. It is the network that performs. It is the app that opens. For it is the unrecognized achievement of any error-free experience. Yet, when stability fails, we are left with few options: try again later or never return. Neither is a preferable user experience.

The overall impact of stability issues is difficult to estimate, but a 2013 Cambridge University study[4] estimated that software bugs alone cost the global economy $312 billion annually. The same study concluded that 50% of all development time was dedicated to resolving bugs. Such costs create a virtual sea of USS Yorktowns, drowning budgets and sinking projects.

Reliability

Treading alongside stability is reliability. *Software Engineering*, by Ian Somerville, states that reliability is "The probability of failure-free operation over a specified time, in a given environment, for a specific purpose.[5]" I like this

[2]US Navy, "USS Yorktown," September 1, 1985.
[3]Seife, Charles. *Zero: The Biography of a Dangerous Idea.* New York: Penguin, 2000.
[4]Comms, Online. "Financial Content: Cambridge University Study States Software Bugs Cost Economy $312 Billion per Year." CJBS Insight. September 08, 2015. Accessed June 15, 2018. https://insight.jbs.cam.ac.uk/2013/financial-content-cambridge-university-study-states-software-bugs-cost-economy-312-billion-per-year/.
[5]Sommerville, Ian. *Software Engineering.* Boston: Pearson/Addison-Wesley, 2004.

definition because it frames reliability in relative terms. Reliability is relative to a specified time. Consider the website-hosting stalwart of "99% uptime." That may sound impressive, until you realize that 99% of a year leaves 87 hours of instability, which is nearly one and a half hours per week. Your personal blog would likely be fine. But, with 99% reliability, Amazon.com would endure a one-billion-dollar loss of net sales. Moreover, reliability is relative to a given environment and purpose. A 99% reliable website hosting may leave its users disappointed. However, 99% reliable SCUBA equipment would literally leave its users breathless.

Problems are inevitable. Errors happen. Apps crash. Sites timeout. We can't plan for every outcome, but we can anticipate and address common issues:

- Sudden outage: Use a monitoring service and be the first to know.

- User frustration: Tweet your awareness of the outage—let users know you know.

- Checkout errors: Set items to out-of-stock shortly before a planned outage.

- Search penalties: Configure a 503 server response, which tells visiting bots that the outage is temporary.

In the end, stability and reliability are not attributes of software, but instead characteristics of an experience. Unstable and unreliable experiences lead to mistrust. Mistrust leads to abandonment. When users abandon, the entire endeavor sinks.

Security

If you placed an overseas call in the 1980s, you may have spoken over the TAT-8 transatlantic cable. It was a first. Never before had fiber optics crossed the Atlantic Ocean. The cable stretched across 3,200 miles of ocean floor, traversing great rift valleys, passing long-forgotten shipwrecks, and weathering undersea storms. TAT-8 was an impressive achievement; yet, it proved to be an insecure one.

Cables started crossing the Atlantic in the mid 1800s, but none were as powerful as TAT-8. The cable could carry thousands of phone calls and millions of data bytes. Some of the earliest Internet messages traveled along it. Despite its role in laying the groundwork for our modern day communications infrastructure, today we remember TAT-8 more for its curious effect on its surrounding ecosystem: sharks treated the multimillion-dollar cable like a chew toy.

Sharks had swum in the oceans for millennia, but they had likely never encountered anything quite like an undersea fiber optic cable. The vast array of digital communication pumping through TAT-8's fiber optic veins generated strong electric fields. Sharks use electrical fields to hone in on prey animals through a process known as electroreception. Even in complete darkness, species such as the lemon shark (see Figure 9-2) can trace the faint bioelectronic signature of its favorite food, the parrotfish. In retrospect, we should not have been surprised that TAT-8's power provided such a culinary attraction. Designers soon learned to shield the cables, blocking TAT-8's electrical fields and securing its data from the powerful jaws of the unwelcome, undersea diners.

Figure 9-2. Lemon shark at the Sydney Aquarium[6]

When TAT-8 was completed, the securing of data was mechanical. Lines could snap. Connections could break. Sharks could chew on the cable, but they did not try to hack it. The millions of data bytes traveling among the connected academic and banking systems could flow unobstructed, relatively safe from manipulation and malfeasance.

---

[6]Patrick Quinn-Graham ,"Negaprion acutidens sydney2", August 16, 2009.

Everything changed in 1988. Using only 99 lines of code, a computer program spread throughout the early Internet. As it replicated itself from machine to machine, the program slowed and crashed networks across the globe. The tiny Morris worm (as it would soon be called) presented a much greater security threat than any 400-pound shark ever could.

In the decades that followed, waves of malware, viruses, and worms exploited both the operating systems of computers as well as the behaviors of users.

Today, all digital experiences are prone to attack. Email, texts, chats, payment gateways, validation, data storage, and others are compromised with troubling regularity. PrivacyRights.org reports that over ten billion records have been breached since 2005. The barely perceptible scent of data travels across the Internet to places we may never have previously imagined. Target's 2013 data breach[7] started with stolen credentials from a heating, ventilation, and air conditioning vendor. Nearly 27 million Department of Veterans Affairs records were stolen[8] from a laptop taken during in a home burglary. Even data disconnected from the Internet can be stolen; Israeli researchers have proven that air-gapped data can be stolen by modulating the sound of a computer's cooling fan and picked up by a nearby phone.[9] Security begets insecurity.

If we realize how insecure our digital experiences are, we might choose to return to the days of telegrams, paper letters, and cash-only transactions. However, even before the digital age, we were not entirely secure. Our lives were beset with wire frauds, postal scams, and strong-arm robberies. We have simply replaced analog insecurity for digital insecurity. In many ways, our collective delusion of security is what keeps technology moving forward.

When designing experiences, our first layer of defense is absence. Information absent from your application is inherently secure. One cannot breach information that does not exist. Do you really need to save users' credit card information? Phone numbers? Postal addresses? User names? Ask yourself, do you even need users to create an account at all? Because, once we obtain information from users, we must treat it like blood in the water.

---

[7]Cheng, Andria. "Target Data Breach Has Lingering Effect on Customer Service, Reputation Scores." Marketwatch. April 02, 2014. Accessed May 28, 2018. http://blogs. marketwatch.com/behindthestorefront/2014/04/02/target-data-breach-has-lingering-effect-on-customer-service-reputation-scores/.
[8]Electronic Privacy Information Center. "EPIC - Veterans Affairs Data Theft." Electronic Privacy Information Center. Accessed June 21, 2018. https://www.epic.org/privacy/vatheft/.
[9]Greenberg, Andy. "This Researcher Steals Data With Noise and Light." Wired. February 07, 2018. Accessed June 21, 2018. https://www.wired.com/story/air-gap-researcher-mordechai-guri/.

Visual design and copywriting can connote security to users. We have all visited websites that did not meet our expectations. Perhaps we noticed a misspelling or a missing image. Maybe the website simply made us feel uneasy. Uneasiness leads to fear. Conversely, we have all visited websites that exceeded our expectations. Perhaps we read a witty bit of copywriting or viewed a gorgeous photo. Maybe the website simply made us feel comfortable. Comfort leads to confidence.

Interaction design affects perceptions of security. Simple form validations, such as a clear indicator for strong passwords, enhance perceived security (see Figure 9-3). Ensuring pages provide adequate confirmation and error messaging show users the application is cognizant of it being used. Consider a typical error message: frequently, applications appear to be just as bewildered by an error as its users are. It is as if a web server said, "Oh my, that was a surprise!" Some error messages may be unavoidable, but we determine their contents. Vague phrasing such as "Something went wrong" does little to assuage the fears of users when submitting their credit card details. This is the consequence of creators wishing to show that errors are rare—so rare that errors surprise even them. Stating "Sorry, your card was declined" tells a user exactly what is going on—no mysteries. As creators, we should treat errors as expected realities, not as inexplicable phenomena shared by users, designers, developers and copywriters alike.

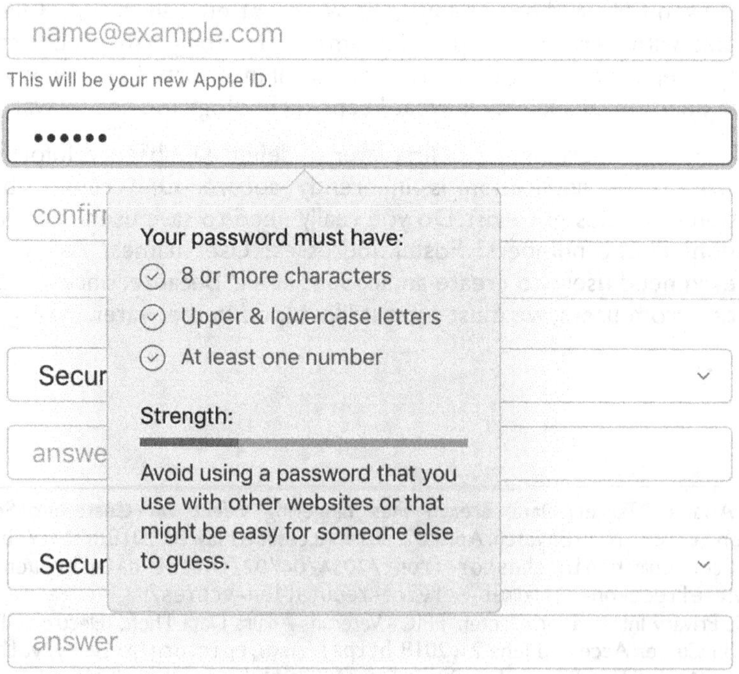

Figure 9-3. Password strength indicator on appleid.apple.com

Lastly, consider the experience of security researchers—the people who uncover vulnerabilities in the products we create. Make it easy for researchers to report their findings; set up a dedicated email address. Be respectful and open-minded in terms of what you hear. No one likes to learn of their own weaknesses. Yet, a tiny indignity received today can save you from a horrific attack suffered tomorrow.

Security is not so much the absence of risk, but the confident acceptance of it. Security is a fundamental requirement for any experience. Fear leads to abandonment. Confidence leads to exploration. Users wade into the murky waters of the unknown and discover what lies beneath the surface of our creations.

Key Takeaways

- All digital experiences can be attacked.
- Insecurity leads to fear. Fear leads to abandonment.
- Comfort leads to confidence. Confidence leads to exploration.
- Do not ask users for unnecessary information.
- Set up a dedicated email address for security researchers.

Questions to Ask Yourself

- What more can I do to provide users a safe and secure experience?
- Do I really need to save user information?
- Do users really need to create an account?
- Did I run a spellcheck?
- Did I correct all obvious visual design bugs, such as broken images?
- Am I requiring users to follow good security practices, such as create strong passwords?
- Have I accounted for all errors that may occur within an experience?
- How can I make it easy for security researchers to contact me?
- What if all my users' private information becomes public?
- What if I am being hacked right now?

Speed

If you want to understand speed, live in the Midwest. It is in our blood. From the earliest age, Ohioans learn that going anywhere else takes a while. Eight hours to New York. Thirty-two hours to California. The Buckeye State is lovely, with its wide prairie lands and lush forests. However, families go elsewhere to vacation. Many fly south to Florida. Others head southeast, toward the Carolinas. My family did neither. We drove to Missouri. Our semi-annual, 10-hour road trip served as a curriculum for a future UX designer, including several lessons about distance, duration, and speed.

We counted cows. We drank Capri Suns. We played Mad Libs—ever play it before? One person reads aloud from a small book containing partially completed sentences. The other players listen and attempt to fill in the gaps See Figure 10-1 as an example.

I went to [**choose a famous location**] and met a [**choose a profession**].

He gave me [**choose a type of food**]. We ate it while [**pick a verb ending in -ing**] the world.

Figure 10-1. Mockup of a make-believe Mad Libs question

One person calls out to the group "I need a famous location... I need a name of a profession... I need a type of food." He or she then fills in the omitted words.

© Edward Stull 2018
E. Stull, *UX Fundamentals for Non-UX Professionals*,
https://doi.org/10.1007/978-1-4842-3811-0_10

The only goal is to create a story. Unsurprisingly, kids try to think of clever phrases (see Figure 10-2).

I went to [**the White House**] and met a [**politician**].

He gave me [**Arby's**]. We ate it while [**destroying**] the world.

Figure 10-2. Mockup of a make-believe Mad Libs answer

Ask. Think. Answer. Repeat. Before you notice, you arrive in Missouri.

For a game first published in 1958, Mad Libs is surprisingly similar to the input-output (IO) mechanisms of modern day software. Software requires specific inputs, such as a first name. Once you input this information, the system might output, "Hello, Bob." With Mad Libs, your input are phrases, and the output is a story. With software, your input is data, and the output is an experience. Speed affects both the input and output.

Imagine the game of Mad Libs, but slow down the process of asking and answering each question by a few seconds, like the following:

"I need a famous location"

...

...

...

"Hmm... let me think about that one..."

...

...

...

...

...

...

"Okay, give me a second; I'm writing that down "

...

"I need a girl's name"

...

...

...

...

...

"Hmm... let me think about that one..."

...

...

...

...

...

...

"Okay, give me a second; I'm writing that down..."

...

...

...

...

...

...

"I need a type of food..." and so on...

Even in this written example, you notice how the slowdown in speed can affect the game. The questions come less frequently. Answers are few and far between. The experience lumbers along, zapping the fun out of each subsequent round of game play.

Speed affects everything. When software is slow, simple tasks frustrate and complex tasks fail.

In Stoyan Stefanov's book, *Book of Speed,* he notes that website users perceive page times to be 15% slower than their actual speed during use. Even more interesting, users perceive the same website to be 35% slower than the actual speed after using it. Our memories of speed are even slower than the actual speed.

We should consider both current and future perceptions of speed when designing experiences. Not only do we want new users, but we also want past users to return. Attracting users is a challenge. Retaining them is a triumph. Those users could have gone jogging, grabbed a bite to eat, played with their kids, walked their dog, done their laundry, watched TV, played a video game, read a newspaper, wrote an email, talked on the phone, got an oil change, shaved their legs, trimmed a ficus tree, danced in their kitchen, read a book about user experience, or really anything. But instead, they visited your website. Yet, after three seconds of waiting for your page to load, more than half of them will have abandoned it.[1]

Mobile users are even less compromising.[2] Mobile websites frequently have only one chance to get it right or suffer the lasting consequences. A full third of mobile users report speed as their largest frustration, and half of those frustrated users will never return.

Mozilla, the makers of the browser Firefox, reduced its page load times by 2.2 seconds and they increased downloads of its browser by 60 million.[3]

Amazon.com famously noted that a decrease of 100 milliseconds increases their overall revenue by 1%.[4] Consider that for a moment. Amazon net sales were $107 billion dollars in 2015. One percent of that is over 1 billion dollars. So, one-tenth of a second could buy three Boeing 777 jetliners.

A user's perception of speed is the result of expectation minus duration. The shorter the expectation, the quicker an experience must perform. Patience becomes a vanishing commodity—not unlike the users themselves.

The Hick-Hyman Law

You witness the Hick-Hyman Law (or Hick's Law) every day. You select a pair of socks, you peruse the aisle of a grocery store, you tap a link within a website menu. You might be surprised to learn that researchers have studied the speed of such interactions for the better part of a century.

The experimental psychologist, William Hick, published his groundbreaking research "On the Rate of Gain of Information[5]" in 1952. He measured the response times of study participants when confronted with multiple choices.

---

[1] "Mobile Speed Impacts Publisher Revenue - DoubleClick." Google. September 2016. Accessed June 07, 2018. https://www.doubleclickbygoogle.com/articles/mobile-speed-matters/.
[2] "Speed Is Key: Optimize Your Mobile Experience." Google. September 2015. Accessed June 07, 2018. https://www.thinkwithgoogle.com/marketing-resources/experience-design/speed-is-key-optimize-your-mobile-experience/.
[3] Cutler, Blake. "Firefox & Page Load Speed – Part II." The Mozilla Blog. April 5, 2010. Accessed June 07, 2018. https://blog.mozilla.org/metrics/2010/04/05/firefox-page-load-speed-%E2%80%93-part-ii/.
[4] Linden, Greg. "Marissa Mayer at Web 2.0." Early Amazon: Auctions. November 9, 2006. Accessed June 07, 2018. http://glinden.blogspot.com/2006/11/marissa-mayer-at-web-20.html.
[5] Hick, W. E. "On the Rate of Gain of Information." *Quarterly Journal of Experimental Psychology* 4, no. 1 (1952): 11-26. doi:10.1080/17470215208416600.

Participants were shown a series of lights. A lightbulb flashed, then a participant pressed a corresponding button. A moment later, another lightbulb flashed, then a participant pressed its corresponding button. Researchers would measure a participant's response times from viewing a flash to selecting a button.

The study's results showed correlations between the number of lights and response time. The more choices a person must consider, the longer a person takes.

Flashing lights and pressing buttons may not rival Nintendo's *The Legend of Zelda* or Microsoft Excel, but the study can tell us a lot about human-computer interaction.

On the surface, Hick's Law proves the old adage "less is more": two choices are better than three; one alternative is better than two. However, that is an oversimplification. Having more choices leads to longer reaction times, but other factors also affect a person's decisions. We remember. We practice. We reason.

Subsequent studies concluded that participants quicken their reaction times through practice.[6] Your interaction with system controls—radio buttons, check boxes, list menus, and the like—also improves over time. As the saying goes, "How do you get to Carnegie Hall? Practice, practice, practice."

Additionally, we must consider a user's prior knowledge. A menu may contain a listing of 50 states, but a user already knows where she lives. Such thoughts are nearly instantaneous. A user also makes selections based on the alphabetization of a list (e.g., she looks for "N" because she wants to select "New York").

Related studies have shown that people sometimes slow down their review of a shorter list and speed up their review of a longer list. Content plays a role. You would likely review a list of former lovers more attentively than a long list of auto parts. Unsurprisingly, longer lists generate more recall errors, while shorter lists generate fewer. (No offense to your love life intended.) A short list is easier to remember, but it is not always faster to review.

Hick's Law offers us a practical insight: we need to balance the number of choices with the speed of making a choice. Choices should illuminate the user experience, not snuff it out. After all, user experience is more than a flashing bulb.

[6]Schneider, Darryl W., and John R. Anderson. "A Memory-based Model of Hick's Law." *Cognitive Psychology* 62, no. 3 (2011): 193-222. doi:10.1016/j.cogpsych.2010.11.001.

Key Takeaways

- When an experience is slow, simple tasks frustrate and complex tasks fail.

- Our memories of speed are slower than the actual speed.

- A user's perception of speed is the result of expectation minus duration.

- The more choices a user must consider, the longer a user takes to consider.

- Users quicken their reaction times through practice.

- Users sometimes slow down their review of a shorter list and speed up their review of a longer list.

Questions to Ask Yourself

- How does an experience perform using a low-bandwidth network, such as 3G?

- When do users expect an experience to begin and complete?

- What can I do to further optimize an experience?

- How much practice have these users had?

- How many choices am I asking the user to consider?

Usefulness

In 1975, the Pet Rock was born. The egg-shaped, smooth stone came nestled in hay, accompanied by an instruction manual, and encased within a small cardboard box (see Figure 11-1). For $3.95, you could own one too. Although the Pet Rock started as a joke, it demonstrates why some products succeed and others fail.

Figure 11-1. Artist's rendering of the Pet Rock with its accompanying straw bed

The Pet Rock's inventor, Gary Dahl, knew it served no practical purpose other than humorous, geologic companionship. Despite the Pet Rock's relative

© Edward Stull 2018
E. Stull, *UX Fundamentals for Non-UX Professionals*,
https://doi.org/10.1007/978-1-4842-3811-0_11

pointlessness, it was sold to over 1.5 million questionably proud owners.[1] People placed them on desks, gave them as gifts, and wrote about them in books. Years later, the website ThinkGeek improved upon the original idea by adding a USB cable connector. And, true to form, the update did not distract its customers with any apparent benefits. How does such a product succeed?

If there is any truth in product development, it would be that a good product fulfills a need, even if that need is entirely made up and supplied to the consumer. Advertising creates a hole, then a product fills it. Want to feel pretty? Buy shampoo. Dead-end job got you down? Eat at Chili's. Trapped in a loveless marriage? Buy a bow-wrapped BMW. Products fulfill needs—real or imagined—in consumers' minds. The Pet Rock succeeded because it, too, fulfilled a need, albeit a silly one. It highlighted the vacuousness of products that do not serve a purpose. People thought such an observation was funny and innovative. The Pet Rock got lucky. Few products succeed in such a way. As creators, we are tasked with a heavier burden. We must design satisfying experiences that fulfill users' needs. We cannot create Pet Rocks.

Do the experiences you design fulfill a need? This fulfillment could be as specific as managing thermostat settings, or as general as entertaining children. Experiences are not inherently satisfying. Just ask anyone sitting in traffic or aimlessly reviewing his or her Twitter feed.

To some extent, all designed experiences attempt to satisfy users with information, entertainment, and capabilities. Some do it better than others. The *New York Times'* website transforms complex subjects into comprehensible stories and engaging media, allowing its users to better understand their world. Consider the website's 2016 Webby Award-winning article, "Greenland Is Melting Away" (https://goo.gl/gSYFWp). It not only informs, but it also entertains and educates users with a richly visual display. It shows global warming's impact, starting with an aerial view of Greenland's southeastern coast, zooming down to a scientific basecamp sitting on an ice sheet.

The game *Minecraft* teaches visuospatial reasoning and goal-setting strategies, engrossing its users in environments that they themselves create. The game not only entertains, but it also informs and enables players to construct and experiment. From the time a player starts the game, she may wander a landscape full of resources that can be used to build structures, maintain crops, and collaborate with other players.

---

[1]Fox, Margalit. "Gary Dahl, Inventor of the Pet Rock, Dies at 78." *The New York Times.* March 31, 2015. Accessed June 07, 2018. https://www.nytimes.com/2015/04/01/us/gary-dahl-inventor-of-the-pet-rock-dies-at-78.html.

Microsoft Excel arranges large datasets and calculations, enabling millions of managers, strategists, and number-crunchers across the globe. With over 750 million users, Excel's usage is ubiquitous. Its user experience may be a diamond in the rough, but you cannot knock its utility. The spreadsheet application not only enables users with capabilities, but also informs—and for a few weirdos, it even entertains. You only need to talk to someone managing a fantasy sports team to discover Excel's potential (see Figure 11-2).

Team	RS	RA	ACT RS	ACT RA	1 RS	1 RA	2 RS	2 RA		RS #	RA #
Chicago Cubs			0	0	748.44	623.7	749	643		4.62	3.85
Los Angeles Dodgers			0	0	698.22	605.88	712	593		4.31	3.74
Boston Red Sox			0	0	766.26	669.06	735	671		4.73	4.13
Cleveland Indians			0	0	688.5	667.72	716	613		4.25	4.06
Washington Nationals			0	0	707.94	643.14	688	633		4.37	3.97
New York Mets			0	0	669.06	636.66	676	592		4.13	3.93
Houston Astros			0	0	737.1	691.74	749	683		4.55	4.27
Tampa Bay Rays			0	0	677.16	665.82	713	622		4.18	4.11
San Francisco Giants			0	0	654.48	617.22	644	595		4.04	3.81
Toronto Blue Jays			0	0	766.26	741.96	765	711		4.73	4.58
New York Yankees			0	0	712.8	677.16	725	686		4.4	4.18
Seattle Mariners			0	0	694.98	672.3	689	662		4.29	4.15
Pittsburgh Pirates			0	0	680.4	656.1	683	667		4.2	4.05
St. Louis Cardinals			0	0	683.64	660.96	671	665		4.22	4.08
Chicago White Sox			0	0	712.8	711.18	701	686		4.4	4.39
Detroit Tigers			0	0	753.3	750.06	692	692		4.65	4.63
Texas Rangers			0	0	743.88	740.34	731	743		4.59	4.57
Arizona Diamondbacks			0	0	672.3	685.26	657	683		4.15	4.23
Miami Marlins			0	0	667.44	670.68	643	685		4.12	4.14
Minnesota Twins			0	0	717.66	745.2	683	707		4.43	4.6
Los Angeles Angels			0	0	696.6	699.84	674	727		4.3	4.32
Kansas City Royals			0	0	691.74	711.18	641	687		4.27	4.39
Oakland A's			0	0	691.74	711.18	668	723		4.27	4.39
San Diego Padres			0	0	605.88	664.2	649	692		3.74	4.1
Baltimore Orioles			0	0	732.24	766.84	697	792		4.52	4.67
Colorado Rockies			0	0	720.9	785.7	680	749		4.45	4.85
Cincinnati Reds			0	0	648	722.52	675	743		4	4.46
Milwaukee Brewers			0	0	687.72	767.88	687	725		4.06	4.74
Atlanta Braves			0	0	609.12	727.38	602	731		3.76	4.49
Philadelphia Phillies			0	0	602.64	735.48	611	721		3.72	4.54

Figure 11-2. An Excel spreadsheet containing a detailed delineation of a fantasy baseball season

The more a product provides relevant information, capabilities, and entertainment, the more satisfying it becomes. An ordinary experience grows into a towering achievement. Make usefulness its cornerstone.

Key Takeaways

- Experiences are not inherently satisfying.

- Satisfying experiences fulfill users' needs.

- Increase the usefulness of products by providing users relevant information, capabilities, and entertainment.

Questions to Ask Yourself

- Do the experiences I design fulfill a user's need?
- How relevant is a particular piece of information to users?
- How can I offer more capabilities to users?
- How can I make an experience more entertaining?

The Lives in Front of Interfaces

Let me share some of my background with you. I started thinking about writing this book several years ago. At the time, I was working at a startup. The company's flagship product allowed employees to manage their healthcare and insurance benefits. It had a sophisticated backend, yet the application was unattractive and difficult to use.

The insurance sector is not known for being a bastion of excitement. Compliance, legal, and regulatory issues mire your design work. Add an unhealthy dose of industry-speak, and you get a prescription for a bad user experience.

The application posed an intriguing set of challenges. Our design updates to the application could not move too fast, lest we upset thousands of existing clients. So, the team made lots of small, frequent changes.

After several months of incremental changes, we arrived at a seemingly innocuous page. Users could review or edit their life insurance benefits. It was unremarkable in every way, containing all the excitement and grandeur you might expect from an insurance form. The team and I rewrote a couple of labels, tweaked a handful of inputs, and called it a day.

© Edward Stull 2018

E. Stull, *UX Fundamentals for Non-UX Professionals*,
https://doi.org/10.1007/978-1-4842-3811-0_12

Jumping ahead a few years, I found myself at a funeral remembering that innocuous life insurance form. I got a sinking feeling, because I had neglected to realize an important fact: the users of that form were not reviewing it to appreciate design or prose. They were there for one reason alone—because a loved one had died. After all, it was a form to review life insurance benefits. They would have likely arrived at the page to review the insurance coverage of their recently deceased spouse or child.

I contemplated a husband or wife struggling through our complicated and convoluted interfaces, searching for information and eventually landing on a cold, procedural-looking web page, all the while being flooded with feelings of sadness and loss. While I had not caused his or her sadness, I certainly was not helping relieve it, either.

The realization embarrassed me, but it also helped me recognize that user experience is about what happens in front of a screen, not within it. We design experiences for other human beings; the software is fine on its own—even bad software. As designers, we ask for users' time, attention and energy, so we must repay them with a good experience.

Being Human

For about 30 minutes, you existed as a single cell. Nine months later, you were born with over 300 bones in your body; yet, you will die with 206.[1] Twenty-two of these form your skull and contain twelve paired cranial nerves, the second being your optic nerve. At just one millimeter wide, it connects your brain to everything visual in the world. On a clear night, your eye can detect a candle's flickering light 30 miles away.[2] Your sense of smell is no less impressive; the human nose can detect a single drop of perfume in an adjacent room, as well as thousands of scents.[3] Each of your fingertips can touch your thumb, and you blush—both of which are unique traits among primates. You hiccup for no apparent reason, as well. Excepting those with physical and cognitive disabilities, people universally laugh, cry, sit, stand, look, touch, and feel. We have been this way for thousands of years. It is both exhilarating and humbling to realize just how similar we all are.

With humans sharing so much in common, you might wonder why designing for them can be such a challenge. After all, you are one of them, too. Knowing what users want and how they behave would seem to be implicitly understood. If that were the case, we would not even need to concern ourselves with user experience. We would know everything already. However, despite our similar physiology and neurology, human beings do have differences.

[1]"List of Bones of the Human Skeleton." Wikipedia. June 07, 2018. Accessed June 08, 2018. https://en.wikipedia.org/wiki/List_of_bones_of_the_human_skeleton.
[2]ArXiv, Emerging Technology from the."How Far Can the Human Eye See a Candle Flame?" *MIT Technology Review*. July 31, 2015. Accessed June 08, 2018. https://www.technologyreview.com/s/539826/how-far-can-the-human-eye-see-a-candle-flame/.
[3]Gerkin, Richard C., and Jason B. Castro. "The Number of Olfactory Stimuli That Humans Can Discriminate Is Still Unknown." ELife. July 07, 2015. Accessed June 08, 2018. doi:10.7554/eLife.08127.

People differentiate themselves through an expansive range of cultures, educations, aptitudes, social norms, etiquettes, and taboos. Some distinctions are subtle, but others are not. A newborn Bulgarian baby may be spat upon for good luck, whereas a Finnish baby may spend its first few nights sleeping in a government-supplied cardboard box.[4] Members of the South American Yanomami tribe eat the ashes of their dead relatives, whereas the Houston-based company, Celestis, launches your loved ones' ashes into outer space (see Figure II-1). Although we have similar starting points, where we go from there often takes wildly different directions.

Figure II-1. On May 22, 2012, a Falcon 9 rocket (similar to the one in photo) carried ashes of 308 people into space[5]

---

[4]Lee, Helena. "Why Finnish Babies Sleep in Cardboard Boxes." BBC News. June 04, 2013. Accessed June 08, 2018. https://www.bbc.com/news/magazine-22751415.
[5]Photo by SpaceX, "Bangabandhu Satellite-1 Mission."

We have similarities. We have differences. Designing a single, optimum experience to serve everyone is impossible; you'd be too busy handing out wet wipes and keeping everyone from launching one another into orbit. Nevertheless, we realize design solutions when we focus our efforts on a particular set of human beings with a particular set of goals.

In this section of the book, we examine the physiological and psychological factors to consider when designing experiences. We discuss how we sense and perceive our world, how our attentions wander, how we are pushed and pulled by persuasion, and how users recognize information today and recall it tomorrow. We even talk about the benefits of being lazy. But, for now, let us be ambitious and start at the very beginning of what makes an experience—perception.

Perception

The English activist and poet John Milton once wrote, "The mind is its own place, and in it self, can make a Heav'n of Hell, a Hell of Heav'n."[1] He composed these words in his epic poem *Paradise Lost*. Through his writings, Milton gave us one of the finest observations about human perception, as well as UX: what we experience is what we perceive.

Paradise Lost is an unsurprising title when you consider what happened in the year it was published. In 1667, Dutch ships sailed up the English River Thames, bombarded towns, burned a dozen vessels, stole the English flagship, and towed it back home to set it up as a tourist attraction.[2] These were the days of unparalleled turmoil. In less than a decade, three civil wars shifted the balance from monarchy to commonwealth to protectorate. Religious strife bled into the politics and political turmoil, igniting religious fervor across England, Scotland, and Ireland. The king dissolved one parliament and fought another. In turn, parliament executed one king and exiled a second. You could not blame Milton for his relativism. He witnessed war, as well as peace; corruption, as well as charity; depression, as well as prosperity. He was also blind.

Milton did not see the world; however, he certainly perceived it. Our senses are only one way we perceive. We sense the world through our eyes, ears, nose, mouth, and skin. We see a fire's flame, hear its roar, smell and taste its smoke, and feel its warmth. Our minds shape these senses into the perception of fire. Psychologists call this bottom-up processing. Conversely, our minds

---

[1] Hammond, Paul. *Milton's Complex Words: Essays on the Conceptual Structure of Paradise Lost*. Oxford: Oxford University Press, 2018.
[2] "Raid on the Medway." Wikipedia. June 19, 2018. Accessed June 21, 2018. https://en.wikipedia.org/wiki/Raid:on_the_Medway.

© Edward Stull 2018
E. Stull, *UX Fundamentals for Non-UX Professionals*,
https://doi.org/10.1007/978-1-4842-3811-0_13

also form perceptions based on prior experiences, general concepts, and expectations. We can look up to the heavens and see stars twinkle against an endlessly black sky. Some of us will see constellations of gods, animals, and objects based on our prior exposure to similar patterns. Psychologists call this top-down processing.

What remains is a perception. Senses transform into fires. Stars transform into gods. Likewise, pixels and screen layouts, beeps and button clicks, swipes and finger gestures, help form our perceptions of digital applications. A dramatic visual design connotes excitement, whereas a zany sound may imply childlike wonder. Your expectations of a game app may align with such senses although a spreadsheet application may not. This is our perceptual processing at work. Understanding how perception works helps improve user experiences. It makes bad software better. It makes good software great. After all, it can make heaven out of hell.

Top-Down Processing

The Viking 1 Orbiter, a NASA spacecraft, took a curious photo when it flew past the Martian surface in 1976. The image, taken by its two vidicon tube cameras, showed a landscape pockmarked with craters and shadowed by mesas. The mostly featureless region of Mars named Cydonia was nothing extraordinary, excepting the humanoid face staring up from its surface. First popularized by the book, *The Monuments of Mars: A City on the Edge of Forever,* by Richard C. Hoagland, this image became known as the "Face of Mars." The covers of grocery aisle tabloids have celebrated the image ever since.

Based on the pixelated 1976 image, the face appears to some viewers as an expressionless, somewhat androgynous, forward-facing human portrait (see Figure 13-1). They believe the photo shows an elaborate construction built to welcome or warn curious onlookers as they fly past the Red Planet.

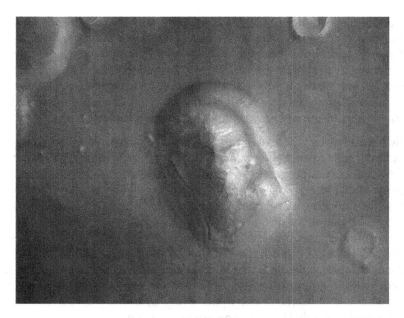

Figure 13-1. Mars' Cydonia region[3]

In actuality, the "Face of Mars" is a rather ordinary mesa reaching 800 feet high.[4] And, like any tall thing under sunlight, it casts shadows. When photographed, the broad shadows it created looked like a jawline, deep shadows looked like eye sockets, and small shadows looked like a nose.

Believing the "Face of Mars" is an alien glamour shot or a garden-variety rock pile is a secondary issue. The primary question is why do most of us see it as a face? Our answer lies in a discussion of schemas.

A schema is a mental shortcut, a way to interpret incomplete information. Say you are gazing up at the stars on a clear evening. You marvel at the expanse of the universe and see a flickering light. For a fleeting moment, you witness a few dim flashes streak across the dark canvas of the night sky. You might think it is a star, a plane, or a UFO. However, there is a good chance that the dim flashes are a passing satellite. Many people are unaware that you can view satellites, including the International Space Station, with the naked eye.[5] Our preconception—our schema—for a light in the sky doesn't include

---

[3]NASA, JPL, and Univ. of Arizona. "Popular Landform in Cydonia Region." Digital image. NASA Image and Video Library. April 11, 2007. Accessed June 6, 2018. `https://images.nasa.gov/details-PIA09654.html`.

[4]NASA. Accessed June 21, 2018. `https://science.nasa.gov/science-news/science-at-nasa/2001/ast24may_1`.

[5]Mathewson, Samantha. "How to Spot the International Space Station with New NASA Tool." Space.com. October 20, 2017. Accessed June 08, 2018. `https://www.space.com/34650-track-astronauts-space-new-interactive-map.html`.

satellites, but we do have a schema for shining stars, passing planes, and even UFOs. Your schema for a star may have started when a parent pointing to the heavens said, "Look sweetie, that's a star." The first time you viewed a plane passing in the night sky, you likely confused it for a star, because your schema for stars was already well established. Later in life, your schema for planes became established, as well. Upon reading this chapter, your schema for unknown lights in the sky now includes satellites, if it didn't already.

People have a strong schema for human faces. We even have a place within our brains dedicated to processing faces, the fusiform face area (FFA). Whereas the brain's visual cortex processes every other visual stimulus (from paper clips to rocket ships), the FFA gains an efficiency through its unique role. A 2009 fMRI study showed that humans can recognize a face in 130 milliseconds,[6] roughly half the amount of time it takes to blink your eye. Studies show that four-month-old infants process faces almost as quickly as adults.[7] The noted scientist, Carl Sagan, hypothesized in his book *The Demon-Haunted World: Science as a Candle in the Dark* that humans evolved a hyper awareness of faces to recognize the emotional states of humans and other animals. The smiling face of a parent indicated a safe opportunity to bond, whereas the snarling face of a predator indicated a strong warning to flee.

From sunup to sundown, our minds sort the world into a series of perceptions. Schemas shape these perceptions, like the gravitational forces of nearby planets, pulling in everything from within their orbits. Some perceptions burn up upon reentry, while others land and become the foundations of new ideas.

Mental Models

Nineteen seventy-seven was a good year. I had just turned seven years old and won a model-building contest at a local hobby store. The prize wasn't money, but fame—the type of fame one might expect to receive from having your work displayed in a suburban Ohio strip mall. Handbills plastered the store's front window. Behind the expanse of glass sat a small wooden platform displaying jet fighters, catapults, Wild West stagecoaches, and my model of a P51 Mustang fighter-bomber (see Figure 13-2). It rested atop a field of green-dyed sawdust grass. Each model paid homage to its source, with every plastic piece glued by hand, every decal affixed by tweezers, and every bolt painstakingly painted with a pin. You might be surprised to learn how many similarities are shared between model building and user experience.

[6]Hadjikhani, Nouchine, Kestutis Kveraga, Paulami Naik, and Seppo P. Ahlfors. "Early (N170) Activation of Face-specific Cortex by Face-like Objects." Advances in Pediatrics. March 04, 2009. Accessed June 08, 2018. https://www.ncbi.nlm.nih.gov/pmc/articles/PMC2713437/.
[7]Farzin, Faraz, Chuan Hou, and Anthony M. Norcia. "Piecing It Together: Infants' Neural Responses to Face and Object Structure." *Journal of Vision*. December 01, 2012. Accessed June 08, 2018. doi:10.1167/12.13.6.

Figure 13-2. P51 Mustang fighter-bomber parked on airfield[8]

In retrospect, the contest afforded the storeowner a free merchandise display; but I received much more in return: for a few days, I mentally transported myself into the seat of a P51 Mustang. Although nearly 40 years have passed, I still remember those daydreams of flying over the countryside of Beavercreek, Ohio.

Surely, we have all had similar experiences when growing up. You imagined being an astronaut, winning Miss America, or living the life of a gunslinger. Yet, just as a nerdy kid like me had no background in flying fighter-bombers, it is likely that you had no direct connection to any of those imagined pursuits. So how could we imagine them so clearly?

We form mental models of perceived experiences. Like flying a toy plane, we construct these models based on our related, past experiences: the books we've read, the movies we've watched, the conversations we've had. We build mental models of software much in the same way.

Google shapes the mental model for search (see Figure 13-3). Likewise, Amazon does for e-commerce; eBay does for auctions; Twitter does for microblogging; and Microsoft Excel does for spreadsheets.

---

[8]VinnyCiro. P-51 fighter. Digital image. Pixabay. January 22, 2015. Accessed June 6, 2018. https://pixabay.com/en/plane-aircraft-military-p-51-607224/.

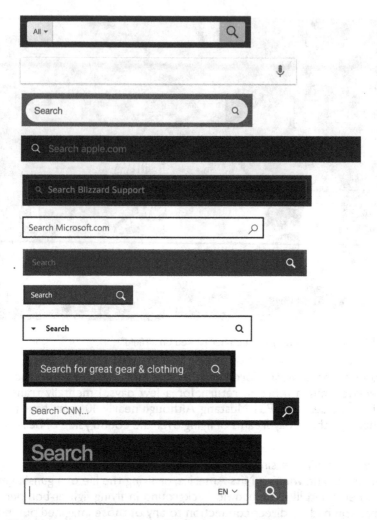

Figure 13-3. Various search controls displayed top to bottom: Amazon.com, Google.com, Target.com, Apple.com, Blizzard.com, Microsoft.com, Playstation.com, Ual.com, Walmart.com, Rei.com, CNN.com, Fox.com, and Wikipedia.com

If a mental model sounds like a schema, take comfort that you are not alone in assuming this. People often confuse the two; however, they differ in a few key ways. Mental models include schemata, but also behaviors and outcomes. Whereas a schema may describe a plane, a mental model describes a plane, as well as flying and landing it.

Mental models set expectations. However, expectations are merely contours that we can sculpt into other forms. New experiences need not be strict combinations of past ones, but mental models are often where new ideas take flight.

Just Noticeable Differences

You are a frog. You sit happily in a cool pot of water. You have not a care in the world. The water starts to warm. It is rather pleasant, reminding you of your time in Palm Springs. The water grows warmer. You look around, admiring the sturdy craftsmanship of your new metal home. The water grows warmer. You relax, reflecting on your life as an amphibian. The water grows warmer. You enjoy your unplanned sauna. The water boils... You are no more.

Why didn't you jump out? One minute you basked in delight; the next minute you boiled in disbelief. You may have escaped if the temperate changes had been more noticeable. Psychophysicists call such a change a "just noticeable difference."[9] A just noticeable difference, or JND, is a unit of measurement. It describes the smallest detectable difference between two levels of stimuli.

Weber's Law

The German professor and experimental psychologist, Ernst Heinrich Weber, first codified JNDs in the 1800s through his research on human touch and sensory physiology.[10] He noted that people notice the relative change in stimuli, not the absolute. He experimented with everything from differently weighted blocks to fluctuating musical notes, but today his law is best demonstrated by your television.

TV commercials often play at a higher volume than the scheduled programming. Your ability to notice this change depends on the volume of your television: low- and medium-volume settings highlight the difference, while high-volume settings mask the difference. You notice the fluctuation between a quiet TV episode and loud commercial because of the relative difference between a low and a high volume is great. You don't notice the fluctuation between a loud TV episode and a slightly louder commercial because the relative difference between the two volumes is small.

Coincidentally, the fluctuating volume between TV programming and commercials was so noticeable that it led to many complaints to the Federal Communications Commission, culminating in the Commercial Advertisement Loudness Mitigation (CALM) Act of 2010.[11] To bypass the legislation, some commercials now fluctuate volumes within themselves to achieve a consistent average volume, while still having the same effect.

---

[9]"Just-noticeable Difference." Wikipedia. May 28, 2018. Accessed June 08, 2018. https://en.wikipedia.org/wiki/Just-noticeable_difference.
[10]"Weber–Fechner Law." Wikipedia. June 06, 2018. Accessed June 08, 2018. https://en.wikipedia.org/wiki/Weber%E2%80%93Fechner_law.
[11]"Loud Commercials." Federal Communications Commission. April 19, 2016. Accessed June 08, 2018. https://www.fcc.gov/media/policy/loud-commercials.

Marketing JNDs

Marketers use JNDs to determine pricing and discounts. Imagine walking past a storefront that offered a "5% Off Sale!" Not only would you continue to walk past the store, but also you would hardly even acknowledge the offer. Retailers often need to move percentage discounts into the 20-25% territory in order to attract the attentions of consumers. Marketers focus on JNDs because they wish to grab your attention without leaving their money on the table. If a consumer responds to a 20% discount, why offer 25%? Doing so would give away the additional 5% for no reason. Some companies—and even entire industries—become trapped by offering inflated JNDs. Discounts of 50%, 60%, and 70% off are viewed with as much skepticism as enthusiasm. When was the last time you saw a furniture store that was *not* having a "Going out of Business Sale"? Because these JNDs are so extreme, the just noticeable difference is barely noticeable at all.

Information Design JNDs

Edward Tufte, noted statistician and Yale professor, further extended JNDs within the study of information design. His "small multiples"[12] demonstrated differences among similar graphs when placed next to one another. When nearly all the information is the same, the differences are easily observed. Tufte's beautifully crafted books include numerous examples of small multiples, ranging from fly-fishing lures to Japanese calligraphy.

So far in this chapter, we have discussed how human beings use past experiences, concepts, and expectations to understand the world. But this is only half of the equation. Top-down mental processing alone can lead us down blind alleys. To understand the other half of the equation, bottom-up processing, we will exchange our blind alley for a dark one.

Bottom-Up Processing

You find yourself walking down an alleyway. It is late. It is dark. You spent the evening with friends at a downtown restaurant. Your car is parked a few streets away. Hulking silhouettes of trash dumpsters and piles of debris line the pathway, as the stench of unfinished entrees and discarded mop water fills the air. You navigate past each starlit obstacle and hear a noise. A footstep. Another follows. You speed up. So do the footsteps. Your shoulders tighten. Your stride widens. Your pulse grows. You gasp and turn and see nothing. It was merely an echo.

---

[12]Tufte, Edward Rolf. *Envisioning Information*. Cheshire, CT: Graphics Press, 2017.

Stories of dark alleys are clichés and staples of horror films. The sights, smells, and sounds lead characters into states of fear and panic. Shadows transform into monsters, stenches instill dread, and unidentified footsteps indicate malicious pursuit. In actuality, the protagonists of such stories should embrace a top-down approach to their thinking: we live in the safest time in all human history, therefore what we sense is rarely danger. The 2014 UN Office on Drugs and Crime report states that we have a 1 in 16,000 chance of being murdered.[13] Considering the average 30-year-old person has greater than one in five chance of living to 100,[14] it would be more rational to worry about your retirement savings, rather meeting your demise in a dark alley.

But our senses can surprise even the most logical of us. Opposed to top-down processing, where we construct perceptions based on general concepts, bottom-up processing constructs perceptions based on sensory data. Listen up. I smell something fishy. This feels squishy. Watch out! Sound, smell, sight, and touch construct perceptions upwards. Sometimes senses aid us; sometimes they trick us. Like victims in a horror movie, we only realize our perceptual errors when it is too late. Yet, we can also leverage bottom-up processing to improve experiences, through which senses become tools to accelerate behaviors, assuage fears, and satisfy users.

Gestalt Grouping

Red represents everything from love to anger, from safety to danger, from Christmas to communism. A red heart shows affection. A red hand shows revolution. A red light tells us to stop. A red exit sign tells us to go. In every case, we sense red in the exact same way. Its wavelength always measures around 650 nanometers.[15] Yet, how we perceive color changes. The conflict between what we sense and what we perceive lies at the core of user experience. We possess a myriad of senses—sight, sound, smell, taste, touch, temperature, and pressure. However, our perception of these senses is often a matter of gestalt.

A favorite word bandied about at dinner parties and in film schools, gestalt describes the whole of something, as opposed to its parts. When you watch a movie, you see action as a continuous motion. One frame at a time, movie stars dangle each other off the bows of ships, avatars fly among floating mountains,

---

[13]"Dicing with Death." The Economist. April 12, 2014. Accessed June 08, 2018. https://www.economist.com/international/2014/04/12/dicing-with-death.
[14]"How Likely Are You to Live to 100? Get the Full Data." The Guardian. August 04, 2011. Accessed June 08, 2018. https://www.theguardian.com/news/datablog/2011/aug/04/live-to-100-likely.
[15]"Reading on Color & Light, Part I." Arizona State University. Accessed June 08, 2018. https://www.asu.edu/courses/phs208/patternsbb/PiN/rdg/color/color.shtml.

and toys tell you their stories. You see this motion in 24, 25, or 48 frames per second. Psychologist call this effect "apparent motion." You experience each frame (each part), but you also experience the movement (the whole). You experience a gestalt grouping.

Gestalt grouping laws describe how people perceive objects as organized patterns. The laws cover similarity, proximity, continuity, and closure. Of these, two are especially helpful in user experience: proximity and similarity.

Proximity

We perceive objects placed next to one another as a group. Our ancestors looked into the night sky and found asterisms and constellations—groups of stars resembling eagles, charioteers, crabs, harps, gods, and dragons. We consciously and subconsciously group together the items we see throughout the day and night. Whereas we may intentionally group stars twinkling in the northern sky as the Big Dipper, we unintentionally group strangers walking down a sidewalk as being acquaintances.

The proximity of interface items implies groupings. We group page titles with nearby paragraphs. We group buttons with neighboring forms. Conversely, when we separate items spatially, we also separate them perceptually. Consider the following example:

Company Name:

First Name:

Last Name:

Is the "First Name" related to the "Company Name" or to the "Last Name"? Due to its proximity, users may associate the first name to the company, rather than last name.

The greatest separation happens when designers scatter information across multiple screens. Doing so destroys gestalt groupings. A designer may forget that only she knows what appears next within an application. If a designer describes something behind a curtain, it will remain a mystery to users until that curtain is opened. A designer has prior knowledge; a user does not. Users need to see information themselves, be it a confirmation page, an error screen, or the next step in a process. Unseen information is not information at all. Dense and elegant information can be found in all sorts of applications, from weather apps to tax preparation software. Conversely, sparse and awkward information may be found in even the simplest of experiences. The difference between elegant and awkward is frequently less a matter of what than where.

Similarity

When objects are similar to one another, we often perceive them as a group. The law of similarity affects user experience in several ways. Frequently, similar visual treatment of screen elements implies a common grouping. Consider the following:

Planes Trains Automobiles

This example displays elements in a similar manner, using the same typeface, the same font size, and the same color. We understand planes, trains, and automobiles to all be methods of transportation. The grouping and treatment of these elements is sensible. Let's change that.

Planes **TRAINS** Automobiles

A user would wonder what a designer implies with such a treatment of the word "TRAINS". Is it more or less important than planes and automobiles?

Planes Trains Automobiles Rocket Goats

As you can see, introducing any new element with a similar treatment maintains the group. Rocket goats may not be a method of transportation, but the example implies that they are. Make something look similar to something else, and users will assume that it is.

Pitfalls of Similarity

Imagine a mobile app with two buttons of similar size and color (see Figure 13-4).

Figure 13-4. Two similar buttons placed in close proximity

The first button, "View Cat Photo," raises no eyebrows; pressing the button might display a photo of a cat. Pressing the second button, "Detonate Explosive," is another story entirely, with presumed dire consequences. You must exercise caution when presenting vastly different behaviors in the same manner. Similarity connotes UX behavior in a linear progression. The first button sets the context for the second button. Viewing cat photos is a safe activity, thereby deescalating the dangerous activity of detonating explosives. To prove this point, let us switch the order of our previous example (see Figure 13-5).

Detonate Explosive   **View Cat Photo**

Figure 13-5. Reversing the order of two similar buttons changes each button's context

Now the button "Detonate Explosive" sets the context for "View Cat Photo." Certainly, this button order makes one pause for at least a moment.

Important behaviors command separation within an interface. Checkout buttons, delete functions, and quit without saving are candidates for such treatment.

Within the larger universe of UX, the Law of Similarity[16] affects behaviors across products and services. Users come to expect similarity. Users anticipate searches to behave like Google (see Figure 13-6) and checkouts to perform like Amazon. Users assume everything from password retrieval, to email unsubscribes, to social shares, will perform in similar ways. We should not design everything to be the same, but we should at least acknowledge users' expectations of sameness. The experiences you design will be compared to many others. And you are outnumbered.

Figure 13-6. Microsoft Bing's search looks and behaves in a remarkably similar way to Google's search

Gestalt grouping transforms random patterns into designed experiences, connecting the detached and highlighting the overlooked. It makes the new feel familiar. Handle gestalt grouping wisely, and you will be in good company.

---

[16]Soegaard, Mads. "The Law of Similarity - Gestalt Principles (1)." The Interaction Design Foundation. May 2018. Accessed June 08, 2018. https://www.interaction-design. org/literature/article/the-law-of-similarity-gestalt-principles-1.

Selective Perception

Hiawatha Service 332 bypasses the steady state of arterial road traffic between Milwaukee and Chicago. In 89 minutes, the train's riders depart Miltown's intermodal gateway and eventually find themselves in the heart of Chicago's Union Station. Long an early morning refuge for blurry-eyed salespeople and late-night party-goers, Hiawatha attracts a wide assortment of professions, cultures, and hangovers. However, each rider's trip is unique, because each is a selective perception.

Commuters and tourists alike fall asleep in peaceful unison within minutes of the train's departure. Gaping mouths and contorted postures fill each carriage like dozens of goldfish placed on blue fabric seats, as the constant hum of the track passes beneath. Dah-dunk. Dah-dunk. Dah-dunk. The rhythm lulls even the most caffeinated to sleep. Some riders doze motionless. Others roll and fidget. They close their eyes. They wear headphones. They curl into the fetal position and form makeshift pillows out of jackets and sweaters. Consciousnesses rise and fall, governed by the lucidity of dreams and the placement of armrests.

Our brains search for stimuli, be it while riding a train or viewing a mobile app. We seek the pleasant and avoid the unpleasant. If our seat is comfortable, we relax and slumber. If light gets in our eyes, we close the shade. Aggravation yields to comfort. Pleasure beats provocation. Social psychologists call this selective exposure, or the "confirmation bias." We find the messages that confirm our beliefs rather than challenge them. We readily notice ads for products we already own. We eagerly recognize virtues in the political candidates we already support. We unhesitatingly accept compliments about things we already enjoy.

One glowing attribute casts a halo around all others. This halo effect affects everything from interpersonal relationships to international branding. We believe attractive people are also kind. We think profitable companies are also managed well. We will even defend our favorite brands by ignoring their competitors' advertising.[17] You can experience this phenomenon firsthand: persuade an iPhone user to switch to Android; convince a Ford truck owner to buy a Chevrolet; coax a Snow user to download Snapchat; cajole a Diet Coke drinker to order a Diet Pepsi. Such halos surround us, enveloping our decisions in predictable delusion.

---

[17]Fennis, Bob M. and Wolfgang Stroebe. *The Psychology of Advertising*. London: Routledge, 2009.

Moreover, we erect psychological barriers to threatening stimuli. Smoking causes nearly one in five deaths.[18] Texting causes one in four car accidents.[19] Unprotected sex causes one in two unplanned pregnancies.[20] Yet, even after being exposed to the dangers of smoking, texting while driving, and unprotected sex, a great number of us still smoke, text, and spend anxious moments awaiting the results of a test strip.

Psychological barriers affect the user experience of digital products, as well. A recent Pew Research report indicates that only 9% of social media users feel very confident that their records are private and secure.[21] However, the user base of such apps continues to grow, totaling 69% of the American public.[22] Users weigh the tradeoffs between privacy and utility, though risks and rewards are often perceived selectively. Likewise, as of January 2017, 4% of Samsung Galaxy Note 7 phones[23]—the defective and recalled devices that may spontaneously combust and burn their owners—have yet to be returned. Few user experiences carry such dire consequences, but we should note that even possible immolation is not a compelling enough argument to offset some people's selective perception.

We avoid the pitfalls of selective perception by acknowledging them. Users can even benefit from their inability to fully perceive an experience, thereby focusing on the necessary.

We can also avoid the pitfalls of selective perception by recognizing its triggers. If you wish to buy a car, your perception will become focused on cars. Recognize that the ads and offers you will find are the result of this trigger, even though the information was always there, waiting for you to perceive it. That car commercial you just watched may be more of a selective perception than a true bargain.

---

[18]"Smoking & Tobacco Use." Centers for Disease Control and Prevention. May 15, 2017. Accessed June 08, 2018. https://www.cdc.gov/tobacco/data_statistics/fact_sheets/health_effects/effects_cig_smoking/index.htm.

[19]National Safety Council. "NSC Releases Latest Injury and Fatality Statistics and Trends."News release, March 25, 2014. National Safety Council. Accessed June 8, 2018. https://www.nsc.org/Portals/0/Documents/NewsDocuments/2014-Press-Release-Archive/3-25-2014-Injury-Facts-release.pdf.

[20]"Gateway to Health Communication & Social Marketing Practice." Centers for Disease Control and Prevention. September 15, 2017. Accessed June 08, 2018. https://www.cdc.gov/healthcommunication/toolstemplates/entertainmented/tips/UnintendedPregnancy.html.

[21]Rainie, Lee. "Americans' Complicated Feelings about Social Media in an Era of Privacy Concerns." Pew Research Center. March 27, 2018. Accessed June 08, 2018. http://www.pewresearch.org/fact-tank/2018/03/27/americans-complicated-feelings-about-social-media-in-an-era-of-privacy-concerns/.

[22]"Social Media Fact Sheet." Pew Research Center: Internet, Science & Tech. February 05, 2018. Accessed June 08, 2018. .

[23]Dolcourt, Jessica. "Samsung Galaxy Note 7 Recall: Everything You Still Need to Know about What's Coming next." CNET. January 23, 2017. Accessed June 08, 2018. https://www.cnet.com/news/samsung-galaxy-note-7-return-exchange-faq/.

As designers, we can intercept these triggers, thereby directing users to what is necessary. For example, imagine someone shopping for a train ticket. He wants to find the best price. His focus is on the ticket's cost, perceiving it above all other stimuli. He may not notice a train's departure time. Placing vital information, such as the departure time, next to the price helps the user avoid a costly mistake. He catches the error before it happens, because we anticipated his selective perception.

Like daybreak filtering through a window, perception highlights some stimuli while obscuring others. It sifts through countless sights, sounds, smells, tastes, and touches, transforming what we could experience into what we will experience. Perception shapes our world. Patterns emerge, messages take form, and users awaken.

Key Takeaways

- Top-down processing occurs when people form perceptions based on prior experiences, general concepts, and expectations.

- Bottom-up processing occurs when people form perceptions through sensory data.

- Understanding how perception works helps improve user experiences.

- A schema is a mental shortcut, a way to interpret incomplete information.

- People have a strong schema for human faces.

- Mental models include schemata, behaviors, and outcomes.

- People form mental models based on our related, past experiences.

- JNDs (just noticeable differences) describe the smallest detectable difference between two levels of stimuli.

- Marketers use JNDs to determine pricing and discounts.

- Small multiples demonstrate differences among similar items.

- People notice the relative change in stimuli, not the absolute.

- Gestalt grouping laws describe how people perceive objects as organized patterns.

- We perceive objects placed next to one another as a group.

- When objects are similar to one another, we often perceive them as a group.

- Selective perception leads people to seek the pleasant and avoid the unpleasant.

- Halo effects influence our perception by assigning the value of a known attribute to an unknown attribute.

- People erect psychological barriers to threatening stimuli.

- We can also avoid the pitfalls of selective perception by acknowledging it and recognizing its triggers.

Questions to Ask Yourself

- Have I considered users' existing mental models?

- What gestalt groupings have I intentionally and accidentally created within the experience?

- Does the placement of items within my design connote a relationship where none exists?

- Does the similarity of items within my design connote a relationship where none exists?

- Will users recognize critical information within the experience?

- Are there any areas of the experience that users may find unpleasant or fearsome?

Attention

In the Midwest, you can find hiking trails that are open all year long. Ohio offers some of the finest. Scorching summer months give way to crisp falls. As winter unfolds, tree-lined rivers freeze into slabs etched by the footprints of thrill-seeking deer. Early spring is lovely. Once barren timber thickets explode with dense foliage and wildflowers, creating an eight-million-acre salad of black walnut, white ash, purple coneflowers, and other ingredients, the most unique being a fruit tree called the pawpaw.

When I first learned of the pawpaw, I thought it was a trick played on gullible hikers. First, the tree had a silly name, sounding like a backwoods euphemism. Second, the pawpaw's description was hard to believe: its fruit weighed up to two pounds, tasted somewhat like a banana, and grew on a tropical-looking tree (see Figure 14-1). Although Ohio offers many things, none of it could easily be described as "tropical-looking."

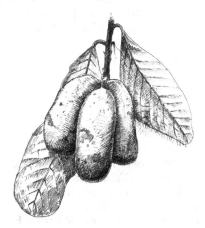

Figure 14-1. Artist's rendering of pawpaw fruit

© Edward Stull 2018
E. Stull, *UX Fundamentals for Non-UX Professionals*,
https://doi.org/10.1007/978-1-4842-3811-0_14

A quick Google search confirmed that pawpaw were real. Then I found one. Now, I find them all the time. Despite hundreds of hours hiking in the Midwest, I had never noticed the odd fruit tree growing off-trail. Though the pawpaw were always there, I was blind to their existence.

Inattention Blindness

In the 1970s, an American psychologist named Ulric Neisser conducted a study on why people sometimes overlook easily seen information.[1] Psychologists refer to this phenomenon as inattention blindness or selective attention. Researchers have replicated Neisser's study numerous times, but Harvard University's variation[2] is the most well known. Christopher Chabris and Dan Simons asked the study participants to watch a video showing basketball players passing a ball and to count the number of passes made by players wearing white.

Please watch the video before reading further. You can view it here: https://goo.gl/L7RRWo.

The video is a little over one minute long and ends with no apparent fanfare. Half of the study participants did not notice anything unusual with the video; they watched players passing a basketball. The other half were surprised to see a gorilla stroll across the screen. Were you? When people pay attention to one activity—be it counting basketball passes or submitting an online form—they are less likely to notice anything else.

Our minds act like a sieve, filtering an endless stream of sights and sounds. One filter is frequency. Along with Neisser's observations, another research study by Dr. Andrew Bellenkes revealed that people are less able to notice something when it happens infrequently.[3] You may be surprised to learn what people tend to overlook when using digital products. Instructions sit unread. Buttons lay untapped. Links remain unfound. Gorillas stroll by unnoticed. Although such failures complicate digital experiences, they are only a small part of a much larger set of processes.

---

[1] Carpenter, Siri. "Sights Unseen." Monitor on Psychology. April 2001. Accessed June 08, 2018. http://www.apa.org/monitor/apr01/blindness.aspx.

[2] Chabris, Christopher F., and Daniel J. Simons. *The Invisible Gorilla: Thinking Clearly in a World of Illusions.* London: HarperCollinsPublishers, 2010.

[3] Bellenkes, Andrew H., Christopher D. Wickens, and Arthur F. Kramer. "Visual Scanning and Pilot Expertise: The Role of Attentional Flexibility and Mental Model Development." Exercise and Sport Sciences Reviews. Accessed June 08, 2018. https://experts.illinois.edu/en/publications/visual-scanning-and-pilot-expertise-the-role-of-attentional-flexi.

Automatic and Controlled Processing

When we talk about attention, we describe how our minds process information. Our mental processing is both automatic and controlled.

Automatic processing handles routine and predictable tasks, such as driving down an open highway. The environment whizzes past us, never requiring us to take much notice. Trees. Fields. Cows. More trees. More fields. More cows. Minutes fly by with hardly a passing thought; yet our minds keep the car on the road, like an invisible chauffeur.

Controlled processing commands more mental resources. If automatic processing is akin to driving a car, controlled processing is akin to texting on a cell phone. We can do either. However, if we do both at the same time, the consequences can prove disastrous. To avoid a sudden obstacle while driving, such as a child crossing the road, requires additional mental resources. They short circuit processing and transform an innocuous text message into a roadside tragedy. Your chances of being in a car accident multiply fourfold if you are driving and texting, which is the equivalent to driving with a .08 blood alcohol level, according to the National Highway Traffic Safety Administration.[4] We all must steer clear of the collisions between controlled and automatic processing.

You employ both forms of attention when using an application. Familiar experiences become increasingly more automatic. Think about how you zoom past Terms of Service agreements. People rarely read them. A 2008 Carnegie Mellon University study indicates that we see nearly 1,500 of these agreements each year.[5] Yet, such agreements can describe practices ranging from hacking indemnification to recording your phone calls. Like trees and cows whizzing past our car windows, so goes our security and privacy.

Unfamiliar experiences require controlled attention. Navigating a new application can feel a lot like a treasure hunt. You search for products. You consider a purchase. You add to your cart. Over time, some of these actions become automatic through repetition, but complex actions require some controlled processing regardless of how many times you do them. Many e-commerce websites have nearly perfected the transformation from controlled to automatic processing. Websites, such as Amazon.com, remove persistent navigation to reduce the user's controlled processing needs. After all, one person's controlled processing is another person's sale.

---

[4]Chase, J.D. Catherine. "U.S. State and Federal Laws Targeting Distracted Driving." Advances in Pediatrics. March 2014. Accessed June 08, 2018. https://www.ncbi.nlm.nih.gov/pmc/articles/PMC4001667/.
[5]Madrigal, Alexis C. "Reading the Privacy Policies You Encounter in a Year Would Take 76 Work Days." *The Atlantic*. March 01, 2012. Accessed June 08, 2018. https://www.theatlantic.com/technology/archive/2012/03/reading-the-privacy-policies-you-encounter-in-a-year-would-take-76-work-days/253851/.

Stroop Effect

When you read this sentence, contrasting lines visualize within your brain's occipital lobe. The back half of its cortex recognizes individual letterforms and pattern combinations of letters. In near parallel, predictions and recognition of words generate within areas of your limbic system and frontal and temporal lobes. Areas with names more fitting for microbreweries than neuroanatomy, such as Wernicke's, Broca's, and Geschwind's, take over and determine meaning and pronunciation. At any stage in this process, new information may intercede, and your coherent thought can be lost in a hazy static.

You expend cognitive resources when viewing information. When something confuses you, the cost is even higher. Your reaction time slows, which a Stroop test proves. Developed in the late 1920s by the American experimental psychologist J. Ridley Stroop, the Stroop test[6] demonstrates the cognitive interference caused by competing stimuli. Consider the following example. Read aloud the TEXT COLOR of the following words:

1. Black   2. White

You likely said the text color of item #1 was black, and hesitated on item #2. Both lines of text use black typography; however, item #2 reads, "White". Competing stimuli can cause a momentary dissonance for users, akin to a cognitive highway pileup. If two words and one color can confuse, imagine the possibilities of a software interface, with its multitude of buttons, links, images, and words. We are lucky that users have short attention spans.

Attention Span

Attention span fluctuates widely, based on age, culture, and context. Several studies estimate that sustained attention lasts for approximately 10 minutes.[7] You demonstrate sustained attention when listening to a lecture, reading a book, or watching a movie. Focused attention is fleeting—sometimes it lasts several minutes, sometimes only a few seconds. Hearing an email alert diverts us momentarily; replying to an email diverts us considerably. After too many diversions, and our flow is interrupted.

---

[6]Stroop, J. Ridley. "Stroop Color Interference Test." *PsycTESTS Dataset*, 1935. doi:10.1037/t31299-000.
[7]Weinschenk, Susan. *100 Things Every Presenter Needs to Know about People*. Berkeley, CA: New Riders, 2012.

Key Takeaways

- Inattention blindness causes users to sometimes overlook easily seen information.

- Users are less able to notice something when it happens infrequently.

- Automatic processing handles routine and predictable tasks.

- Controlled processes command a user's attention.

- Unfamiliar experiences and complex actions require controlled attention.

- Users expend cognitive resources when viewing information.

- The Stroop effect results when competing stimuli cause a cognitive interference.

- Attention span fluctuates based on a user's age, culture, and context.

- Users' sustained attention lasts for approximately 10 minutes.

Questions to Ask Yourself

- What do my users see but do not notice?

- How frequently does an event occur within an experience?

- What about an experience is routine and predictable?

- What about an experience requires focused attention from my users?

- What else competes for my users attentions?

- Do users view any contradictory information within an experience?

- What is the expected duration of an experience?

- How often is an experience interrupted?

- Am I unnecessarily interrupting users?

Flow

You may not think of yourself as a gamer, but you have likely heard of Pac-Man. Namco's 1980 hit video game has appeared everywhere from Atari 2600 to Windows 10. Take a moment and picture the game in your mind's eye. See its maze-like screen. Hear the telltale sound of "waka waka waka" as you maneuver Pac-Man through sharp turns and long straightaways. You avoid multi-colored ghosts. You seek flashy power-ups. You cheer for 10,000-point bonuses. Each game level consists of many tiny experiences connected along a circuitous path. Do you recall what Pac-Man eats along this path? Gold coins.

Gold coins motivate players of Pac-Man, as well as users of countless other experiences—both digital and analog. A well-known gem of writing advice is to place gold coins in your work. Readers pick up and examine these momentary scenes, curious quotes, and bits of dialog. A trail of gold coins encourages people to keep reading, moving readers from one part of a story to the next. For example, when writing this chapter I found a funny anecdote about Pac-Man:

> Pac-man was originally named Puck Man. The game's American manufacturer, Midway Games, changed the name to Pac-Man to prevent vandalism to the game's coin-operated cabinet. Cabinets were placed in video arcades and emblazoned with big, bright game logos. As you might imagine, a mischievous teen with a marker could easily change a "P" to an "F".

© Edward Stull 2018
E. Stull, *UX Fundamentals for Non-UX Professionals*,
https://doi.org/10.1007/978-1-4842-3811-0_15

If you found the anecdote sufficiently interesting, perhaps you will continue reading the remainder of this chapter. Likewise, gold coins can be used when designing all sorts of experiences.

A Google search for Pac-Man returns several gold coins (see Figure 15-1). Within the deluge of 25,400,000 results, you find information about Pac-Man dolls, Pac-Man wallpaper, and Pac-Man swimwear. You see Top Stories, ranging from a Beatles parody of Pac-Man, to Namco President Masaya Nakamura's obituary, to even a Pac-Man Doodle game by Google. These bits of information pique your interest and keep your attention until you find your desired result. Google knows a user's journey may begin and end with a simple search. However, more importantly, Google knows its lifelong relationship with a user depends on a series of repeated, connected experiences. Gold coins make such experiences flow.

Figure 15-1. A Google search for "pacman" returns a full range of both ordinary and fascinating results. The latter serves as enticements for further discovery (i.e., gold coins).[1]

---

[1]Google search results for "pacman". Digital image. Google Search. Accessed June 07, 2018. https://www.google.com/search?q=pacman.

Mihaly Csikszentmihalyi wrote in his book, *Flow: The Psychology of Optimal Experience*, that flowing experiences immerse people in focused attention. Flow is "being in the zone." We may feel flow when playing a game, searching the Internet, writing a story, running a marathon, or doing countless other activities. In a flow state, our attention is balanced between arousal and control. An activity provides enough of a challenge to maintain our interest without it overwhelming our abilities.

Video games challenge us to pursue extraordinary goals. Defend planets from attacking aliens. Beat waves of falling tetrominoes. Rescue princesses from barrel-throwing gorillas. But they also encourage us to complete smaller, interconnected activities. We avoid asteroids, rotate shapes, and leap over barrels. Each of these smaller activities fit within our available attention spans. Attempt to fit too many and our attention wanes. Somewhat like an Internet connection, our attention is throttled by an approximate 110-bits-per-second bandwidth.[2] We use this bandwidth to leap from one small activity to the next. When these leaps become seamless, our entire experience flows. Thoughts clear and time shrinks.

Flow alters time. Long periods of time shorten. Video game players spend hours chasing high scores, building characters, and conquering worlds. Players may become hyperfocused, going without food or rest. In 2015, a Taiwanese man died during a three-day gaming binge at an Internet cafe.[3] Perhaps more tellingly, his body sat motionless for over four hours before it was discovered, his death going completely unnoticed by his fellow players. Even the flow of casual games alters our perception of time. Pokémon Go players play for an average of 45 minutes per day.[4] Thank goodness the duration is no longer, or else our sidewalks would be littered with mobile phones and former seekers of Pikachu.

Compared to games, business applications carry far less risk of hyperfocus. Excel crunches numbers. Slack manages messages. Photoshop edits images. Such applications help people do work. Some do it better than others. However, we rarely use such applications for the sake of pure enjoyment. Even a poorly designed business application can succeed in the marketplace if no alternatives exist. Yet, once users find a better way to accomplish their

---

[2]Csikszentmihalyi, Mihaly. "Transcript of "Flow, the Secret to Happiness"." TED: Ideas worth Spreading. February 2004. Accessed June 08, 2018. https://www.ted.com/talks/mihaly_csikszentmihalyi_on_flow/transcript.
[3]Hun, Katie, and Naomi Ng. "Man Dies after 3-day Internet Gaming Binge." CNN. January 19, 2015. Accessed June 08, 2018. https://www.cnn.com/2015/01/19/world/taiwan-gamer-death/index.html.
[4]Dogtiev, Artyom. "Pokémon GO Revenue and Usage Statistics (2017)." Business of Apps. May 4, 2018. Accessed June 08, 2018. http://www.businessofapps.com/data/pokemon-go-statistics/.

goals, lackluster applications become distant memories. Business applications must flow to survive. Once-dominant applications are overshadowed by their nimbler, more-focused, better flowing rivals. Consider the tectonic shifts in software, where huge companies such as the 15,000-person Adobe now find themselves competing with the likes of the 29-person Bohemian BV, the makers of Sketch. When users seek better experiences, companies suffer. Lotus. Netscape. Palm. Users will abandon companies without hesitation, discarding years of their design and development efforts with a mere tap of a Quit button.

We can learn much from games. For a game to be successful, players must choose it from among thousands of choices, dictated neither by necessity nor utility. Aliens won't actually attack Earth. Tetrominoes won't actually fall from the sky. Princesses won't actually be stuck in their castles forever. Players invest their time, money, and energy for no better reason than to have an enjoyable experience. As creators of experiences, we should ask our users for nothing more.

Key Takeaways

- Place gold coins within your work to maintain user attention.

- Relationships with users depend a series of repeated, connected experiences.

- In a flow state, our attention is balanced between arousal and control.

- Flow states alter time; long periods of time shorten.

- Business applications must flow to remain competitive.

Questions to Ask Yourself

- Where within an experience do users get bored, distracted, or overwhelmed?

- Where can I add smaller, easily achievable goals?

- Do users maintain their interests and control within an experience?

- Am I protecting users' wellbeing and safety throughout an experience?

- Do users have a more enjoyable alternative?

Laziness

You can learn a lot about design from a mountain goat. More specifically, you could learn a lot about design by watching how a mountain goat moves. Despite being mountaintop daredevils perched on precarious peaks and straddling rocky heights, they do not in fact climb mountains (see Figure 16-1). They walk, run, and skip up mountains. Jumping from boulder to outcrop at heights sometimes reaching hundreds of feet, a mountain goat searches for food and shelter in among the bluffs. However, their fearless activities do not highlight the mountain goat's most important quality: they are lazy.

Figure 16-1. Wild goats, such as the Alpine ibex, inhabit mountain ranges across Europe[1]

---

[1]Kaz. Mountain Goat. Digital image. Pixabay. September 16, 2015. Accessed June 7, 2018. https://pixabay.com/en/goat-mountain-mountain-goat-ibex-940896/.

© Edward Stull 2018
E. Stull, *UX Fundamentals for Non-UX Professionals*,
https://doi.org/10.1007/978-1-4842-3811-0_16

Even though appearances might deceive us, mountain goats do not randomly choose their path up a rocky mountainside. They look for whatever foothold is easiest to reach. A mountain goat doesn't debate the inherent dichotomies of risk and reward, or the perceived benefits of a complicated approach; no, a mountain goat chooses its path by whatever seems to be the quickest way up—the path of least resistance.

In terms of user experience, creating a path of least resistance is a virtue. Laziness is often maligned, as if it were an attribute limited to the uninspired and neglectful. Yet, laziness is also a matter of efficiency: it represents the least effort necessary to achieve a goal—and nothing more (see Figure 16-2). We witness such efficiencies in everything from economics to linguistics. The principle of least effort, first conceived in 1894, forms the basis for modern information science. People seek information using the most convenient, fastest method available. They stop looking once they find a minimally acceptable result. This behavior is what psychologist Daniel Kahneman described as "System 1" in his seminal book, *Thinking Fast & Slow*. Only after exhausting convenience and immediacy will people engage in higher-order, analytical thinking—otherwise called "System 2."

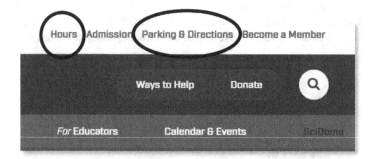

Figure 16-2. Providing direct links to key screens (e.g., Hours and Parking & Directions on a museum website) saves visitors' time and alleviates potential frustrations[2]

Apply this concept to your own online activities. Do you want to fill out form fields? Watch compulsory ads? Receive convoluted driving directions? These questions are of course rhetorical. Such experiences disorient users and foster their abandonment. Instead, we must provide an unobstructed ascent to their goals. A completed goal is the pinnacle of user experience. Favor laziness and you will conquer human nature, as well.

---

[2]The Works Utility Navigation. Digital image. The Works. Accessed June 7, 2018. https://attheworks.org/.

Key Takeaways

- People seek information using the most convenient, fastest method available.

- Remove obstacles within an experience to hasten the completion of user goals.

Questions to Ask Yourself

- How can I provide additional convenience to users?

- What is the fastest conceivable means for users to reach their goals?

Memory

In September 1959, psychology researchers Lloyd and Margaret Peterson tested how quickly people forget information.[1] The researchers asked their study participants to remember a random trigram, a three-letter group (e.g., KHZ). Next, the researchers gave the participants a three-digit number (e.g., 375). The participants were then instructed to count backward from the number by subtracting three (e.g., 372, 369, 366, etc.).

Try it yourself. Ready?

1.   Remember "KHZ".

2.   Set a timer for 12 seconds.

3.   Count backward by three, starting with 375.

4.   Look away from the page and return to it after the timer completes.

I'll wait.

Do you recall the trigram? If you did, pat yourself on the back. If you did not, take comfort in the fact that few of the original study's participants did either.

On a combined average, the study's results showed that after six seconds, only 50% of the letters could be remembered. After 12 seconds, only 15%. The test seemed simple, but something interfered.

---

[1] McLeod, Saul. "Saul McLeod." Simply Psychology. January 01, 1970. Accessed June 08, 2018. https://www.simplypsychology.org/peterson-peterson.html.

© Edward Stull 2018
E. Stull, *UX Fundamentals for Non-UX Professionals*,
https://doi.org/10.1007/978-1-4842-3811-0_17

Counting down by three is a classic example of interference. The counting behavior creates new information in our minds—the number was 375; minus 3; number is now 372; minus 3, number is now 369; and so on. New information overshadows past information, diminishing our ability to store and retrieve memories. Try as you might, your short-term memory becomes exhausted.

The same interference affects user experience. For example, we must remember all sorts of information when shopping online: pricing, sizes, scores, reviews, reward points, availability, and many others. Taken individually, each bit of information is easy to recall. However, we also perform multiple behaviors when shopping online: navigate between screens, add to shopping carts, create passwords for accounts, and input shipping information into forms. Each behavior may interfere with our memory.

If you need users to remember something, such as a price, keep it displayed on the screen, because out of sight really is out of mind. Were those red shoes $20 or were the blue ones? Did I ship that purchase to my old or new address? Did I use Mastercard or American Express? As the experiment showed, even three letters can be difficult to recall. When you ask users to remember, you must remember that they will forget.

Recognition Trumps Free Recall

Do you remember "pop quizzes" during your days of elementary school? It was no wonder that we all preferred multiple-choice over fill-in-the-blank answers. Recognizing data is frequently easier than recalling data without order or context (i.e., free recall).

When designing applications, our users recognize and recall memories from an entire range of experiences. Words, pictures, icons, and functionality have a context in both your application and every other experience. Users assume that "Cancel" means an application will stop an operation. They anticipate a backward arrow "<" to return to a previous screen. They expect a website's checkout to work like other checkouts. They recognize rather than recall.

Explicit and Implicit Memory

Memories you consciously recall are explicit. Like a customer placing a lunch order, you make a request for a specific memory. You attempt to remember if you jogged yesterday. "Brain, tell me if I jogged yesterday," you ask. "Coming right up!" your brain replies. Your brain usually delivers the order. Explicit memories can often be sequenced, such as verbalized into a story. You woke up. You then drank some coffee. Afterward, you jogged while listening to your favorite song.

You recall explicit memories when using software: "What was my password?", "How do I crop this image?", "How do I invite my friends to a Facebook event?" Your brain's hippocampus plays an important role in explicit memory, consolidating information experienced throughout your day.

Implicit memory is trickier. You do not order up implicit memory; it is akin to unconsciously remembering how to use a fork while eating. You just seem to know it. You have repeated the activity so many times that the behavior seems ingrained. Depending on your cultural background and where you have lived in the world, you may use chopsticks in much the same way. To an average American, these eating utensils may feel awkward and clumsy. Yet, to a frequent user of chopsticks, they are both practical and effortless.

Overtime, the repeated recall of an explicit memory may lead it to become an implicit memory. You utilize implicit memories when using software, too. You once learned how to bold a selection of text, but you likely now do it automatically. Many keyboard commands become automatic. Saving a document becomes only a matter of telling yourself you need to save—not the explicit memory of how to save. Your fingers magically align themselves on your keyboard. Voila! You save your work. Your brain's amygdala plays an important role in implicit memory.

Researchers studied the relationship between the hippocampus and amygdala in a fascinating experiment[2] involving three participants, a flashing blue light, and a loud boat horn. Study participants included a person with a damaged hippocampus, a person with a damaged amygdala, and a person with a damaged hippocampus and a damaged amygdala. The study was simple: When a blue light flashed, researchers blared an unpleasantly loud boat horn into the study participants' ears. Blue flash. BAHH-ROOOOO! Blue flash. BAHH-ROOOOO! Blue flash. BAHH-ROOOOO! Afterward, the researchers asked each participant about the event to check their explicit memories.

The person with the damaged hippocampus did not remember that the flash and sound happened at the same time. The person with the damaged amygdala did remember.

The researchers then checked the participants' implicit memory. The blue light flashed again. Flash. The person with a damaged hippocampus reacted to the flashing light, even though he did not remember the accompanying boat horn. However, the person with the damaged amygdala did not react to the flashing light, even though she remembered the boat horn. The third person did not remember or react to either the light or the horn.

Explicit memory requires the brain's hippocampus. Implicit memory requires the brain's amygdala.

---

[2]Hall, Richard H. *Explicit and Implicit Memory*. PDF. University of Missouri, 1998. http://web.mst.edu/~rhall/neuroscience/06_complex_learning/explicit_implicit.pdf

Schemata

Think of a bicycle. I bet you can quickly recall many of its parts: wheels, handlebars, pedals, and the seat. Furthermore, you can easy drill down through these memories and recall smaller parts, such as the spokes, the gear switcher, and the seat post. Your memory of the spokes is with your "wheel scheme." Wheels are within your "bicycle scheme." Our ability to quickly retrieve information from long-term memory increases when placed within a schema. You quickly reference this type of memory through repeated exposure. Imagine how easy it would be to recall the parts of a bike, if you repaired bikes for a living. Experts in a field (such as a bike mechanic) can quickly retrieve memories from a single, larger schema, rather than multiple disconnected schemata.

Serial Position Effects

Serial position effects affect a viewer's ability to recall information. You can remember the first few items in a list more accurately than the items in other positions. This is the primacy effect (see Figure 17-1). Conversely, the recency effect allows you to more easily recall the last few items viewed in a list (see Figure 17-2). It is often best to place high-value items at the beginning or end of a list. Users have a hard time recalling the nebulous middle positions.

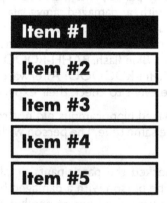

Figure 17-1. Example of the primacy effect

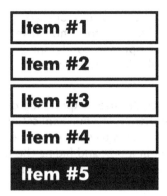

Figure 17-2. Example of the recency effect

How many items can a person easily recall? The answer depends on how the list is "chunked." Nelson Cowan is the Curators' Professor of Psychology at the University of Missouri. His 2010 study[3] on working memory estimates that humans are capable of remembering three to five chunks of short-term memory tasks. An example of a chunk is a letter, digit, or word. These chunks are affected by the length of list items and other factors, such as the age of the test subject. However, the limit of four chunks is generally accepted.

Rewards, Restrictions, and Memory

The American psychologist, Edward Thorndike, ran a series of studies at Columbia University in the early 1900s pertaining to how rewards and restrictions improve or decrease an animal's ability to complete subsequent tasks.

In this particular case, the animal was a cat, and the task was for the cat to escape the confines of a box. The box was opened by an escape lever that could be operated by the cat. If you have a cat, you can empathize with the plight of a cat involuntarily confined within a box. Getting my two cats into their cat carriers nearly requires an act of wizardry.

The cats in this study[4] were rewarded by a piece of fish placed outside the confining box. Thorndike recorded the time it took for the cats to discover how to operate the lever, then recorded the time it took upon subsequent

---

[3]Cowan, Nelson. "The Magical Mystery Four: How Is Working Memory Capacity Limited, and Why?" National Center for Biotechnology Information. February 01, 2010. Accessed June 08, 2018. doi:10.1177/0963721409359277.

[4]"Law of Effect." Wikipedia. April 10, 2018. Accessed June 08, 2018. https://en.wikipedia.org/wiki/Law_of_effect.

trials. The cats got faster at successfully operating the lever. The observed decrease in time became the basis for Thorndikes's "Law of Effect," which further led to concepts of operant conditioning in behavioral psychology. The key takeaway is that behaviors are enforced by successful outcomes and eroded by unsuccessful attempts.

These same behaviors affect the user experience of applications. Early-stage successes in the user experience of your application foster increased usage, failures do not. Forcing your users to sign up on the first screen? Doing so is akin to trapping your users in a box and hiding the escape lever. Instead, give your users an immediate success and you will both be rewarded.

Cryptomnesia

Memory errors are not limited to users; designers suffer the same. Cryptomnesia is the false belief that something is new, when in actuality it is an unconscious memory. This misattribution, either to oneself from an earlier time or—even worse—remembering somebody else's work as your own, can occur at any time. It is particularly problematic during times of stress.[5] And designing experiences can be as stressful as any other profession.

Design is the realization that your best ideas are actually collaborations. When designing a product, take note of those in the room: the project managers, copywriters, developers, and testers. If you need to validate a future solution, include those collaborators. You might uncover valuable insights that were overlooked when an idea was first formed.

Research is the realization that your best ideas are actually someone else's. If you do competitive research, it is a good practice to review your research work a second time after a first draft is made. You might discover that your idea is one that was unconsciously borrowed from a competitor.

Memory is more flexible and imperfect than any of us would care to admit, but we should not despair. We need not concern ourselves with its faults; instead, we should that understand its pliability and incompleteness are what allow us to experience the new. Memory continually emends and rebuilds our past, yet it also creates a foundation for our future.

---

[5]Kim, Jean, M.D. "Don't Blame Plagiarism on Mental Illness." Psychology Today. October 3, 2014. Accessed June 21, 2018. https://www.psychologytoday.com/us/blog/culture-shrink/201410/dont-blame-plagiarism-mental-illness.

Key Takeaways

- Interference diminishes a users' ability to store and retrieve short-term memories.

- Persist important information within a user's view, because out of sight really is out of mind.

- Recognizing data is frequently easier than recalling data.

- You consciously recall explicit memories (e.g., passwords).

- Implicit memories are unconsciously remembered (e.g., how to use a fork).

- Over time, explicit memories may become implicit memories.

- Schemas aide our ability to quickly retrieve memories.

- The primacy effect helps users recall the first few items in a list.

- The recency effect helps users recall the last few items viewed in a list.

- Behaviors are enforced by successful outcomes and eroded by unsuccessful attempts.

- Memories are imperfect and may change over time.

Questions to Ask Yourself

- Are users required to remember information and perform tasks at the same time?

- Is all necessary information to complete a task directly observable to users?

- What happens if users forget vital information?

- What assumptions do users make about an experience?

- How frequently are users exposed to a piece of information?

- Are users required to recall more than recognize vital information?

- Where within a list does vital information appear—can it be moved to the first or last position?

- Which user behaviors does an experience reinforce or impede?

- How might users' memories about an experience change over time?

Rationalization

Days before the French president's death, hours before the New Year, moments before a napkin was removed from his face, François Mitterrand ate two birds whole—bones, beaks, and all.

If you ever see an old photo of restaurant diners wearing napkins over their faces, you are witnessing the consumption of rare, endangered ortolan bunting birds. They are smallish songbirds, each weighing less than an ounce (see Figure 18-1). Skinny and rather plain-looking, the birds would not seem to be obvious targets for gourmands. But, in a true act of culinary barbarism, the ortolans are first force-fed, then drowned in brandy, roasted, and served whole. French custom dictates that a napkin be worn over the diner's face, because he or she ingests the bird like a harbor seal guzzling down a herring.

Figure 18-1. Female Ortolan bunting bird. Drawing by Wilhelm von Wright (1810-1887)[1]

---

[1]Drawing by Wilhelm von Wright (1810 - 1887). Scanned from *Svenska fåglar, efter naturen och på sten ritade* 2nd ed., public domain.

E. Stull, *UX Fundamentals for Non-UX Professionals*,
https://doi.org/10.1007/978-1-4842-3811-0_18

The practice has been outlawed since 1999; however, in the closing hours of 1995, Mitterrand ate the ortolans and promptly died eight days later.

Had he lived longer, Mitterrand would have struggled to justify his behavior. The cruelty of drowning birds would have been reason enough to back away from the table. Additionally, his socialist sensibilities must have been vexed by such a meal, featuring a dish so shameful that the napkin was said to hide the guilty from the eyes of God.[2] Yet, Mitterrand ate it anyways. Twice!

Are human beings rational? If we were, then certainly we would not ride motorcycles, wear high-heeled shoes, pay more for brand-name products, flirt with bad boys and bad girls, prefer expensive over inexpensive wines, or eat endangered songbirds, among other things. Although we make some rational decisions, we make far more irrational ones.

We buy an expensive dress because "it is on sale." We play a violent video game all day because we "need to relax." We let children play tackle football, because "exercise is healthy." We make irrational decisions then reframe them as rational. Psychologists call this behavior post-hoc rationalization.

Post-hoc rationalizations shape how we create products. We produce a successful or failed product then retroactively justify why it is so. Process methodologies result from post-hoc rationalizations. Agile. Scrum. Kanban. Lean. Google Design Sprints. DevOps. V-Model. Extreme Programming. Rapid Application Development. Capability Maturity Model Integration. Waterfall. Whatever. Methodologies have their merits, but they pale in comparison to reason. Rational decision-making can make nearly any project successful; a misapplied methodology can turn a golden goose into a vampire bat.

We can avoid post-hoc rationalization's worst effects by recording our reasoning at the time we make a decision. We all rationalize; we need not compound it with forgetfulness. Documents need not be elaborate or exhaustive; a simple annotation will often suffice. As Parnas and Clements' paper, "A Rational Design Process: How and Why to Fake It",[3] argues, the main benefit of documentation is retrospective: it allows future designers to understand not *how* decisions were made but *why*.

Post-Hoc Fallacy

With a post-hoc fallacy, we mistake correlation for causality. Watch people waiting to cross a busy street. A woman presses a button to trigger the

---

[2]"France Bans an Old Culinary Tradition | News | News & Features | Wine Spectator." WineSpectator.com. July 16, 2009. Accessed June 08, 2018. https://www.winespectator.com/magazine/show/id/8222.

[3]Parnas, David Lorge, and Paul C. Clements. "A Rational Design Process: How and Why to Fake It - IEEE Journals & Magazine." Design and Implementation of Autonomous Vehicle Valet Parking System - IEEE Conference Publication. February 1986. Accessed June 08, 2018. doi:10.1109/TSE.1986.6312940.

crosswalk signal. After several seconds, the signal still says, "DONT WALK". She presses the button again, waits a moment, and presses it a third time. The walk signal now displays, "WALK." She may believe it takes three presses of the button to trigger the signal. Though the events seem interrelated, the first button press started a timer. Once the timer expired, the "WALK" signal was displayed. The wait is always the same whether a person presses the button one time or one hundred times. We see the same with elevator and subway door close buttons, fake office thermostat controls, and anywhere else users mistake what they observe to be what is real (see Figure 18-2).

Figure 18-2. Elevator "door close" buttons are often false buttons that have no effect on the door's operation[4]

---

[4]Brett_Hondow. Elevator Buttons. Digital image. Pixabay. January 21, 2014. Accessed June 7, 2018. https://pixabay.com/en/elevator-buttons-elevator-buttons-248639/.

Post-hoc fallacies may occur in marketing, economics, legal systems, and even the user experience of software. Software users frequently engage in post-hoc fallacies. They believe online forms submit faster when they repeatedly mash buttons. They believe comments are most easily read when written in ALL CAPS. They believe computers screens last longer when they use a screen saver. They believe free shipping is actually free.

Although post-hoc rationalizations and fallacies riddle user experience, we need not remedy every misconception. Some may even be beneficial. If we realized our lack of online privacy, we might stop communicating. If we recognized our lack of security, we might stop discovering. If we understood all the challenges of living and working in today's digital world, we might stop advancing all together. A little bit of rationalization can be a good thing, allowing us to move forward and our ideas to take flight.

Key Takeaways

- Post-hoc rationalizations retroactively justify outcomes.

- Post-hoc fallacies are formed when we mistake correlation for causality.

- Rationalization and fallacy can worsen or improve user experiences.

Questions to Ask Yourself

- What post-hoc rationalizations do I and my team make?

- How can I best document a key project decision?

- What post-hoc rationalizations do my users make?

- Does an experience account for users' post-hoc fallacies, such as repeated form submissions?

- How can I ethically leverage user rationalizations to lead users to better outcomes?

Accessibility

Over four million miles of roadway span the United States. A driver in southeastern Florida City, Florida could hop in her car and 50 hours later step out into northwestern Blaine, Washington 3,435 miles away. Despite her long journey, she would have a remarkably consistent drive. She would travel on 12-foot wide interstate lanes constructed of approved asphalts, aggregates, and finishes.[1] She would view highways signs fabricated from specified microprismatic sheeting and retroreflective paints. She would obey a nearly uniform set of traffic laws. She would experience a system that allows widespread access by nearly every shape, size, and make of vehicle, from Mazdas to Maseratis, from semi-trailers to school buses, from tower ladder fire trucks to Harley Davidson Fat Boys.

Imagine if you awoke tomorrow morning and this system had suddenly changed. Highways were six feet wide. Roads were constructed of gooey tar and jagged rocks. Speed limits were written in tiny, white text on light gray backgrounds. Traffic laws were state secrets. This system would no longer support your needs, making your travel both difficult and dangerous. You could no longer easily get to work, visit a grocery store, or reach a hospital. How would your life change?

Although this thought experiment may seem farcical, one out of six Americans faces similar dilemmas every day. They encounter systems that do not support their needs. They confront challenges to get to work, visit grocery stores, reach hospitals, ascend stairs, read books, play games, order takeout, understand conversations, exchange currency, negotiate contracts, download a mobile app, or use an e-commerce website. They have a disability.

---

[1]"Manual on Uniform Traffic Control Devices." Manual on Uniform Traffic Control Devices (MUTCD) - FHWA. Accessed May 28, 2018. http://mutcd.fhwa.dot.gov/.

© Edward Stull 2018
E. Stull, *UX Fundamentals for Non-UX Professionals*,
https://doi.org/10.1007/978-1-4842-3811-0_19

Merriam Webster Dictionary defines a disability as "a physical, mental, cognitive, or developmental condition that impairs, interferes with, or limits a person's ability to engage in certain tasks or actions or participate in typical daily activities and interactions." Seems fitting. However, the activist and disability pioneer Dr. Henry Viscardi described it another way: "There are no disabled people. We are all just temporarily abled."

Viscardi's definition is equally applicable today. We are all a broken bone, damaged DNA strand, or high-grade fever away from being disabled, be it for an afternoon or a lifetime. Disability ranges from a sprained wrist to severe cognitive dysfunction, and includes vision loss, color blindness, deafness, paralysis, scarring, seizures, neurological disorders, speech impediments, dyslexia, ADHD, and social and emotional issues. Even glaring sunlight or a blaring alarm is enough to disable us temporarily. Try viewing a text message when walking out of a dark theater. Try listening to a voice mail when standing in the middle of a rock concert. Disability need not be permanent to be total.

Although the range of disabilities is broad, one thing is consistent—the desire for accessibility. Accessible products and services allow all people to fully enjoy and participate, bypassing limitations and frustrations, transforming dead-ends into on-ramps.

With its quiet simplicity and inclusive design, GOV.UK provides an accessible repository of government services and information to all UK citizens. The website supports a full range of screen readers, screen magnifiers, and speech recognition software. A person may order audio CDs, Braille documents, and large print versions of the website's content. Eschewing heavy-handed visuals for fast-loading pages, the website's user experience excels on both desktop and mobile devices. Clear, contrasting text make reading a breeze. And, perhaps most importantly, the website's designers do not rely upon intuition—they regularly test accessibility with real users, including those who have physical and mental disabilities.

Compared to the public sector, private businesses often fall short when designing accessible websites and apps. Accessibility may be handled only during the last stage of a project, like a steamroller that flattens and smoothes over the most erroneous errors, filling in the potholes of an experience. Creators sometimes misconstrue accessibility as a cost rather seeing it for what it is: a potential profit center.

Accessibility is not altruism. It is instead an acknowledgement that different people have different needs—a foundational concept underpinning both UX and business. A business might go bankrupt if it refused to serve anyone working within the mining, construction, and the manufacturing sectors. Yet, an inaccessible app or website would underserve roughly the same percentage of people: 12.8%[2] of all Americans. For the sake of market share alone, accessibility makes good business sense.

Smart companies serve both the explicit and implicit needs of users. A person can activate Speak Screen narration[3] on Apple iOS devices (see Figure 19-1), vocalizing the text of a website or app. When paired with Bluetooth-enabled hearing aids, a person could hear the narration without broadcasting her interests across a room of strangers.

Figure 19-1. Apple iOS offers users an efficient means to transform on-screen text into audio narrations[4]

---

[2]Data Access and Dissemination Systems (DADS). "Results." American FactFinder. October 05, 2010. Accessed June 08, 2018. https://factfinder.census.gov/faces/tableservices/jsf/pages/productview.xhtml?pid=ACS_15_1YR_S1810&prodType=table.

[3]"Vision Accessibility - IPhone." Apple. Accessed June 08, 2018. https://www.apple.com/accessibility/iphone/vision/.

[4]"Welcome to GOV.UK." GOV.UK. Accessed June 07, 2018. https://www.gov.uk/.

Microsoft Xbox allows players to form an Xbox Club, a group of like-minded players who may or may not share a disability. In addition, players with limited mobility can utilize Xbox's adaptive controllers[5] to enhance their experience through customizable, oversized controls. What better way to spend a lazy afternoon than to pair up with trusted friends and climb the leaderboards of Forza Motorsports?

When we improve accessibility for disabled users, we enhance the lives of all users. OXO Good Grips kitchen tools were designed for people with arthritis,[6] but everyone likes big, easy-to-hold handles. Sidewalk curb cuts gently transition between streets and sidewalks, helping the 6.8 million Americans who use a wheelchair, cane, or crutches[7] (see Figure 19-2). They also help people riding bicycles, parents pushing strollers, and delivery drivers pulling dollies. Closed captioning helps the reported 20% of Americans with hearing loss.[8] It also allows everyone to follow a story in noisy restaurants and airports.

Figure 19-2. A curb cut offers all users a smooth transition from street to sidewalk. The textured ground surface provides a tactile cue to vision-impaired users and improved traction to everyone.[9]

[5]"Xbox Adaptive Controller | Xbox." Xbox.com. Accessed June 08, 2018. https://www.xbox.com/en-US/xbox-one/accessories/controllers/xbox-adaptive-controller.
[6]www.corepublish.no, CorePublish -. "Inclusive Design - a People Centered Strategy for Innovation." Scandic Hotel - Inclusive Design. Accessed June 08, 2018. http://inclusivedesign.no/product-graphic/oxo-good-grips-article29-265.html.
[7]University of California - Disability Statistics Center. "Mobility Device Statistics - United States." Disabled World. March 18, 2015. Accessed June 08, 2018. https://www.disabled-world.com/disability/statistics/mobility-stats.php.
[8]Hearing Loss Association of America. "Basic Facts About Hearing Loss | Hearing Loss Association of America." HLAA Updates. Accessed June 08, 2018. http://www.yourhearingloss.org/content/basic-facts-about-hearing-loss.
[9]Kidthesaurusdotcom. "Bumpy Curb Sidewalk." Digital image. Pixabay. July 3, 2014. Accessed June 7, 2018. https://pixabay.com/en/bumpy-curb-sidewalk-street-padding-381747/.

Accessibility grows increasingly necessary as the digital natives of today become the aging populations of tomorrow. Both deserve a good experience. We all do.

For more information about accessibility and inclusive design, see:

- American with Disabilities Act (ada.gov)
- European Accessibility Act (goo.gl/GaWw1C)
- Inclusive Design at Microsoft (goo.gl/ybPFXP)
- GOV.UK Accessibility Blog (goo.gl/3R8spG)
- Section 508 of the U.S. Rehabilitation Act (section508.gov)
- W3C Web Content Accessibility Guidelines (goo.gl/bKbTfu)

Key Takeaways

- One out of six Americans have a disability.
- Accessibility products and services may increase their market share.
- Accessibility enhances the lives of all users.
- Accessibility needs grow over time.

Questions to Ask Yourself

- Is an experience accessible to users who have a disability?
- How can I include users with disabilities in my project's research, design, and testing efforts?
- How do temporary changes in a user's environment (e.g., bright sun light, loud ambient noise, or reduced mobility) affect his or her experience?

Storytelling

Human beings started talking to one another approximately 100,000 years ago.[1] We have spent all the time since telling stories—heroic struggles, religious divinities, and chickens crossing roads. We have carved stone slabs, inked animal skins, stained papyruses, set lead blocks, banged on typewriters, and tweeted up to 280 characters.

In ancient Greece, nearly 2,300 years ago, Aristotle (see Figure 20-1) and his contemporaries created a system of storytelling called "rhetoric." It not only withstands the test of time but also informs modern day software design. From antiquity to today, we utilize rhetoric for a variety of reasons, the greatest of which is to persuade an audience.

---

[1]Jackendoff, Ray. "FAQ: How Did Language Begin?" Linguistic Society of America. Accessed June 08, 2018. https://www.linguisticsociety.org/resource/faq-how-did-language-begin.

© Edward Stull 2018
E. Stull, *UX Fundamentals for Non-UX Professionals*,
https://doi.org/10.1007/978-1-4842-3811-0_20

Figure 20-1. Artist's rendering of Aristotle sculpture

John Quincy Adams, the sixth president of the United States, spoke about and wrote extensively on persuasive rhetoric (see Figure 20-2). In his book of lectures entitled *The Art of Persuasion*, he wrote that the primary motivation in any debate is "the attainment of good or avoidance of evil." Put in another way, we tell stories to help one another find success and avoid failure.

Figure 20-2. 1795 painting of John Quincy Adams by John Singleton Copley[2]

_____

[2]John Singleton Copley, "John Quincy Adams" (1796), public domain.

In essence, user experience is persuasive rhetoric: I (the application) tell you (the user) that by doing something, you will achieve a goal. Like rhetoric, user experience contains notions of ethos, pathos, and logos.

Ethos

Ethos involves the perceived character of a person, place, or thing. It helps or hinders appeals made by a communicator to an audience. Orators appeal to listeners, writers to readers, and software to users.

What if a uniformed police officer asked you for your driver's license? You would likely quickly comply. What if a man in Bermuda shorts and a beer t-shirt asked you to do the same? Your reaction would be less enthusiastic, if not wholly dismissive.

We excel at judging credibility. In 2013, a team of researchers from the University of Glasgow measured the speed at which we judge credibility[3] and found that it takes less than half a second. In 300 to 500 milliseconds, approximately the speed of a blinking eye, we make our assessment. We form a first impression in less time than it takes to say the words "first impression."

Audiences evaluate the character, credibility, and reputation of the storyteller. The storyteller need not be a person. Sometimes, the storyteller is a product or service.

Consider the last time you made a purchase, shared your email, or tweeted a link to an article. Perhaps you worried about security, privacy, or accuracy. Hesitant to check out? Apprehensive about spam? Concerned about fake news? All are issues of ethos. Ethos affects behaviors in both the real and digital worlds. You would not keep your lifesavings in a boarded-up building, nor would you type your bank account number into a sketchy-looking website.

An experience's ethos may be trustworthy, treacherous, or display any other attribute. Everything about an experience shapes its ethos—be it a police badge or a beer t-shirt, a website checkout or an email sign-up. Each item augments or diminishes. Each tells a story.

Pathos

Commonly recited in poetry and love letters, pathos attempts to create an emotional connection with an audience. Appeals to pathos may be found in the unlikeliest of places, from Viagra advertisements to Facebook's iconic thumbs-up. Humor. Fear. Love. Disgust. Emotion stirs the heart to action.

---

[3]McAleer, Phil, Alexander Todorov, and Pascal 12 Belin. "How Do You Say 'Hello'? Personality Impressions from Brief Novel Voices." PLOS Medicine. March 12, 2014. Accessed June 08, 2018. doi:10.1371/journal.pone.0090779.

Pathos often misses its mark. Maladroit pathos is commonplace. Malnourished pets stare longingly from TV screens and into our living rooms. Soulless stock art images spray across websites like corporate graffiti. Spammy political emails warn us of a dire crisis requiring our donation. Such pathos dull our senses and defeat their own rhetorical arguments.

An authentic appeal to pathos speaks directly to a user's needs. It avoids the trite and the cliché. It can frame happiness as a healthy lifestyle, channel altruism into community involvement, and reshape fear into education. REI's advertising highlights people enjoying active lives, promoting health and an appreciation of the great outdoors. The website VolunteerMatch uses the tagline, "We bring good people & good causes together," connecting spirited volunteers with local nonprofits. PeacePlayers International organizes kid's sports to bridge cultural, religious, and ethnic divides, spanning the globe from Argentina to Tajikistan.

Pathos illuminates a story with emotion by either darkening a problem or highlighting its solution.

Logos

Logos appeals to our logical senses. Of the three rhetorical appeals, logos most closely mirrors user experience. Users seek rationale—facts, features, and functions—for decision making.

Apple says its new MacBook Pro weighs two pounds.[4] REI Adventures lists a Greek vacation for $4,299.[5] Tesla boasts that its Model S can go 335 miles on a single charge.[6] Such facts offer practical information, but they are devoid of emotional engagement.

Within a digital experience, appeals to logos can be nearly endless. For example, a shopping cart may offer users multiple shipping options—one-day, two-day, or no-rush. Shipping options may list estimated delivery dates—tomorrow, the day after, or next week. Estimated delivery dates may link to information about expedited fees, free shipping information, or no-rush discounts. And so on, and so on… each step informs the next, compelling users to gather further evidence.

---

[4]"MacBook." Apple. Accessed June 08, 2018. https://www.apple.com/macbook/.
[5]"Greek Island Hopper: Hike the Cyclades Islands." REI Co-op Journal. Accessed June 08, 2018. https://www.rei.com/adventures/trips/europe/greek-island-hiking.html.
[6]"Model S | Tesla." Tesla, Inc. Accessed June 08, 2018. https://www.tesla.com/models.

Kairos

Dr. BJ Fogg, noted author of *Persuasive Technology,* defines kairos as an event that happens at the exact right moment in time. Like ethos, pathos, and logos, kairos originates from classical rhetoric.

Kairos in software may include hundreds, if not thousands, of events. A mobile app that provides driving directions gives an advanced warning before a turn should be made. If given too soon, the driver turns too early; if given too late, the driver misses the turn. Transferring money online confronts the same challenges. Transfer it too early, and a person may lose interest revenue; transfer it too late, and she may bounce a check. For a person to effectively complete a task, it must occur at the right time—the more precise, the better.

Homer Simpson, of the TV show *The Simpsons,* once invented an "Everything's OK Alarm."[7] The device matched the size, shape, and shrillness of a typical home smoke alarm. It emitted a piercing series of beeps every three seconds to alert nearby listeners that everything was okay. Only when the alarm fell silent did it indicate a problem. We see "Everything's OK Alarms" affixed all over badly designed experiences in the forms of annoying call-outs and unnecessary notifications. Both are failures of timing, alerting users to information they do not currently need. Turn off the "Everything's OK Alarms" and you will immediately improve the user experience of your applications.

A Careful Balance

We may encounter each rhetorical device on its own, but we often see all four acting together. When an experience is credible, users will trust. When it is emotional, users will engage. When it is logical, users will understand. When it is timely, users will act. The best experiences balance ethos, pathos, logos, and kairos, employing just enough of each to help users achieve their goals. To maintain this careful balance, we must arrange.

Inductive and Deductive Arrangement

Since the dawn of storytelling, two ways of organizing stories have existed. The first way, inductive arrangement, is the most common. With inductive arrangement, we make a claim and then support it. For example, "ABC product is the best product ever! Now here is why." The second way is deductive arrangement. With deductive arrangement, we delay making our claim until we offer a supporting narrative. For example, "Here is what makes a great product, yada, yada, yada… Now here is ABC product!"

---

[7]"The Wizard of Evergreen Terrace." Wikisimpsons. Accessed June 08, 2018. https://simpsonswiki.com/wiki/The_Wizard_of_Evergreen_Terrace.

We arrange experiences as well. Inductive arrangements help users do something quickly. For example, users view a hyperlink labeled "Account" and then click it to view their account. Deductive arrangements help users make considered choices. For example, users read the benefits of creating an account then fill out a registration form.

We act then learn, or we learn then act. Whichever approach you take, know that arrangement only sets the stage for a story. You still have to create meaning and translate it to another person. Rhetoric helps facilitate this communication. However, arrangement and rhetoric are only the conveyance of a story. A story does not happen on a page or screen, but instead is experienced within the minds of the audience members. In the end, the story is theirs alone.

Key Takeaways

- In essence, UX is a form of persuasive rhetoric.

- Ethos involves the perceived character of a person, place, or thing.

- Pathos attempts to create an emotional connection.

- Logos appeals to our logical sensibilities.

- Kairos represents the precise right moment of an event.

- User experience may include notions of ethos, pathos, logos, and kairos.

- Storytelling and UX share similar inductive and deductive arrangements.

- With inductive arrangement, we make a claim and then support it later.

- With deductive arrangement, we delay making our claim until we offer a supporting narrative.

- Audiences transform stories into experiences.

Questions to Ask Yourself

- How do my perceptions of a person change based on his or her social status, wealth, and physical appearance?

- How do my perceptions of a product or service change based on its popularity, price, and marketing communications?

- How can I guide users to beneficial goals through persuasive rhetoric?

- What within an experience augments or diminishes its credibility?

- Does an experience connect with users in an authentic and ethical way?

- What evidence can I provide to users?

- Do users receive information when they need it?

Persuasion

With drugstores dotting our neighborhoods like neon-lit aid stations, we sometimes forget that reliable medication was not always so obtainable. In the 1860s, patent medicine blanketed the American West. Cure-alls, such as *Dr. William's Pink Pills for Pale People*[1] made out of Epsom salt and rust, claimed to offer comfort for ailments ranging from "loss of vital forces," St. Vitus's Dance,[2] and "all female weakness"—medically questionable but certainly unenviable afflictions. The era marked a time in American history, as well as the history of persuasion.

The late 1800s was also the golden age of railroads. The tracks of the First Transcontinental Railroad laid across nearly 2,000 miles of plains, hills and mountains stretching from San Francisco, California to Council Bluffs, Iowa. Millions of pounds of iron, steel, and timber, along with the strenuous efforts of thousands of workers, connected the sandy shores of the Pacific to the prairies of America's heartland. Many of the laborers hailed from China.

---

[1]"Pink Pills for Pale People." Kansas Historical Society. Accessed June 08, 2018. https://www.kshs.org/kansapedia/pink-pills-for-pale-people/10240.
[2]Things Worth Knowing. Schenectady, N. Y.: DR. WILLIAMS MEDICINE COMPANY. https://archive.lib.msu.edu/DMC/sliker/msuspcsbs_drwi_drwilliams2/msuspcsbs_drwi_drwilliams2.pdf

Chinese rail workers, like many migrant communities, brought their culture with them, along with their traditional remedies. One of which was a therapeutic oil made from a snake indigenous to the sprawling rice paddies of east-central Asia. Enhydris chinensis, commonly referred to as the Chinese sea snake (see Figure III-1), had long been used to treat sore muscles of the Cantonese. Long days spent chipping away rail paths through Sierra Nevada granite would tire the back of even the strongest worker. Without an aspirin bottle in sight, these workers turned to the remedy they knew best: snake oil.

Figure III-1. Artist's rendering of Chinese sea snake

For a variety of reasons, the term "snake oil" devolved from a traditional medicine, to medical cure-all, to a pejorative catch-all. Xenophobia and estrangement of the Chinese by white westerners certainly played a role in this. However, the sale of snake oil (see Figure III-2) highlights fundamental aspects of human persuasion and motivation. Believing that a flask of oil holds a cure to anything seems a silly, antiquated notion until we realize that bottles of coral calcium and wild yam cream make some of the same claims, even today.

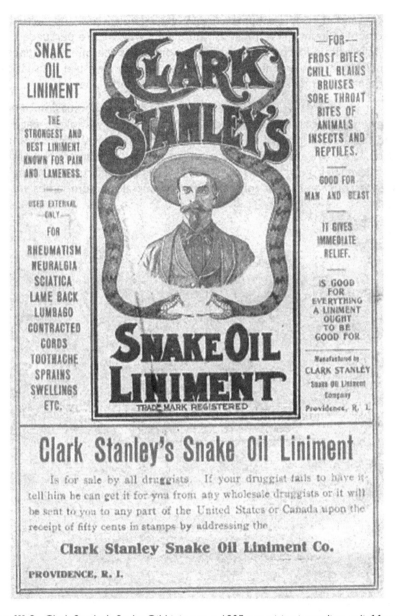

Figure III-2. Clark Stanley's Snake Oil Liniment, c. 1905, promising immediate relief from everything from frost bite to rheumatism[3].

---

[3]"Clark Stanley's Snake Oil Liniment, True Life in the Far West," 200 page pamphlet, illus., Worcester, Massachusetts, c. 1905, public domain.

To be persuaded is to be human. Every generation, class, and culture is a laboratory of influence. We must ask ourselves why we are persuaded. Why do the mysterious forces of desire, selection, and positioning influence our decisions? Moreover, we must ask ourselves how can we use persuasion to help rather than hinder, and guide rather than misguide.

We begin by discussing the fundamental concepts of persuasion: empathy, authority, motivation, relevancy, and reciprocity. Next, we take a page from the book of marketing and cover the "Four Ps": product, price, promotion, and place. Lastly, we answer the questions of why do consumers buy, and how users achieve their goals. In this pursuit, we focus our efforts on crafting good experiences; but, we cannot rely on craftsmanship alone; sometimes we must persuade.

Empathy

President Ronald Reagan was moved. He had just viewed a private screening of The Day After, a 1983 television movie chronicling the aftermath of a nuclear exchange with the USSR. Reagan reportedly[1] developed such a sense of empathy with the fictional, irradiated residents of Missouri that, soon after the screening, he began nuclear disarmament talks with the Soviet leader Mikhail Gorbachev.

A year earlier, the film makers had read Jonathan Schell's, The Fate of the Earth, a book about the consequences of nuclear war.[2] Years before that, Jonathan Schell had read John Hersey's Hiroshima, the Pulitzer prize-winning account of the 1945 atomic bombing of the Japanese city[3] (see Figure 21-1). What resulted was remarkable: the actual deaths of thousands inspired the fictional suffering of millions, which led to the possible salvation of billions. How did these stories connect over such great spans of time and distance? In one word: empathy.

---

[1]Stuever, Hank. "Yes, 'The Day After' Really Was the Profound TV Moment 'The Americans' Makes It out to Be." The Washington Post. May 11, 2016. Accessed June 08, 2018. https://www.washingtonpost.com/news/arts-and-entertainment/wp/2016/05/11/yes-the-day-after-really-was-the-profound-tv-moment-the-americans-makes-it-out-to-be/?utm_term=.91fcefe3d91c.
[2]Fox, Margalit. "Jonathan Schell, 70, Author on War in Vietnam and Nuclear Age, Dies." The New York Times. December 20, 2017. Accessed June 08, 2018. https://www.nytimes.com/2014/03/27/us/jonathan-schell-author-who-explored-war-dies-at-70.html.
[3]Voicesinwartime. "Jonathan Schell in Voices in Wartime - Hiroshima." YouTube. July 20, 2008. Accessed June 08, 2018. https://www.youtube.com/watch?v=mTeVzgkUOLg.

© Edward Stull 2018
E. Stull, UX Fundamentals for Non-UX Professionals,
https://doi.org/10.1007/978-1-4842-3811-0_21

Figure 21-1. The aftermath of the "Little Boy" atomic bombing of Hiroshima, Japan[4]

Empathy frames another person's experience as our own. It extends beyond merely sympathizing with another person; instead, for a brief moment, we see the world through their eyes.

Fostering and building empathy are achieved through two primary means: mirroring and active listening.

Let us start with the easiest one—mirroring. A recent study (Jiang 2012) showed that a part of the human brain, the left inferior frontal cortex (a small region located near your left temple) contains mirror neurons that synchronize between people when they speak face-to-face. The phenomenon may sound like science fiction, but you have witnessed it occur countless times. Try to recall an argument. That back-and-forth you are having in your head is not real. It is you taking the place of the person with whom you were arguing. You might even find yourself inflecting your voice or making hand gestures.

---

[4]US Navy Public Affairs Resources, "Hiroshima Aftermath, cropped version with writing of Paul Tibbets," August 6, 1945, public domain.

Mirroring

You create empathy by first gathering information about a person's perspective and needs. What are this person's hopes, concerns, and fears about a subject? Listen and record the answers. Next, find a spot where you can be alone and repeat back to yourself what you heard. Try to recount the person's exact body language and temperament. Was she hesitant? Was he chortling? Granted, this part of the exercise may seem theatrical, but you are not trying to win an Oscar. Whisper to yourself. You need not yell and swing from the chandeliers. The goal of the exercise is to realize what the person needs—not what you think the person should need. The result is a much richer understanding of the other person's perspective.

Although mirroring may help solidify the bonds between designers and users, it is ineffective in establishing a connection that is being refused. In 2011, a group of researchers from the University of California San Diego conducted a study[5] in which participants watched mock job interviews. Some interviews were straightforward and professional, but others involved the interviewer being intentionally condescending and unfriendly to the interviewee. In each case, the interviewee physically mirrored the interviewer. The study, led by Piotr Winkielman and his colleagues, returned an interesting result: the study participants registered the interviewee as less competent when mirroring an unfriendly interviewer.

The behavior of a hostile person should not be mirrored for many reasons, namely because it fails to build empathy. It is also a scientific argument for keeping your cool; after all, you may occasionally face hostility, but you still have to look at yourself in the mirror each morning.

Active Listening

Active listening is listening without judgment. In this exercise, your job is to understand what you are hearing, not to approve of what you are hearing. Our natural response to hearing another person speak is to form a judgment. When we do, we wait for the person to stop talking, so that we can confirm our own beliefs. A user might tell us, "I think this button should be red. It's my favorite color and… blah… blah…" Then, we begin to think of all the reasons why he is wrong: his failure to notice the blue buttons, his hubris to art direct our work, and his unfortunate hairstyle. Rather than listening, we busy ourselves with what we are going to say in response. Instead, we should

---

[5]Kiderra, Inga. "'Mirroring' Might Reflect Badly on You." Obesity Is 'Socially Contagious,' Study Finds. July 26, 2011. Accessed June 08, 2018. http://ucsdnews.ucsd.edu/archive/newsrel/soc/2011_07mimicry.asp.

focus on the words he is saying. When we begin to form a judgment, we should abandon it immediately. We have the rest of our lives to think about own opinions, but, for right now, at this moment, we want to only listen. Why? Because we cannot empathize with those we do not hear.

Even with empathy, we sometimes confront seemingly insurmountable challenges—challenges that confound, frustrate, and vex the most empathetic of creators. Next, we will tackle wicked problems.

Wicked Problems

Wicked smart. Wicked pissa. Wicked awesome. Wicked old. Wicked muggy. Wicked broke. Wicked stoned. Wicked sad. Wicked good. Wicked fast. Wicked long time. Wicked problems.

If you have spent any time with a native Bostonian, you have undoubtedly heard the word "wicked" used as an adjective, adverb, and a noun. The term is an intensifier: add wicked to any other word and you are left with a greater version of it, albeit one that is occasionally crass. When Horst Rittel and Melvin Webber coined the phrase "wicked problems" in 1973, they too meant to convey a greater version of a word: wicked problems are problems not solvable by reasoning alone.

Reasoning has solved problems in mathematics, chemistry, physics, and biology: Dmitry Mendeleev formulated the periodic table; Grigori Perelman solved the Poincaré conjecture; the CERN team proved the Higgs boson. Elusive as these solutions were to find, right and wrong answers did exist. Right and wrong answers to wicked problems do not.

Without a right or wrong answer, a wicked problem has no endpoint. The wicked problem evolves as more attempts are made to solve it. Social injustice, income disparity, and environmental degradation all stand as examples of wicked problems. They frequently manifest as solutions to other wicked problems. For example, we could eliminate all fossil fuels tomorrow; however, such a solution may harm the economies of developing countries. Even if you disagreed with the premise, you must agree that climate change and economic disparity are difficult problems to solve simultaneously. An answer to one often negates the answer to the other.

Luckily, we face less daunting challenges when designing software. However, we still encounter wicked problems. Goals often conflict: expert users demand flexibility, where novice users desire succinctness; businesses want to sell high, when customers wish to buy low. At best, we address a particular need for a particular person at a particular time. But if there is no right or wrong answer to a wicked problem, how do we design a solution?

Again, we return to empathy. Even the wickedest of problems can be diminished by lessening a problem for just a few people—even one person. Although we never solve a wicked problem, our repeated attempts can meaningfully improve user experiences.

For example, Facebook knows it cannot solve all disputes among its 1.8 billion users. Allowing free-flowing communication while simultaneously preventing objectionable speech is a wicked problem. Improving one complicates the other. However, in 2012, Facebook realized it could improve the experience of users who were being bullied. To do so, Facebook launched its Bullying Prevention Hub.[6] It offered practical advice and useful reporting tools to teens, parents, and educators. Most people will likely never encounter a need for such information, but for the few that do, it could be a lifesaver.

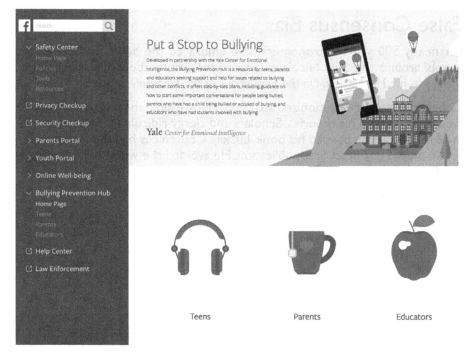

Figure 21-2. Facebook's Bullying Prevention Hub[7]

---

[6]"Put a Stop to Bullying." Dr. Ben Carson's Accomplishments, Awards, Honors, and Community Involvement. Accessed June 08, 2018. https://www.facebook.com/safety/bullying.

[7]"Put a Stop to Bullying." Dr. Ben Carson's Accomplishments, Awards, Honors, and Community Involvement. Accessed June 10, 2018. https://www.facebook.com/safety/bullying.

In an adaptation of the essay *The Star Thrower*,[8] by Loren Eiseley, a boy walked along a sandy shoreline and noticed an old man standing amongst countless beached starfish. The old man picked up the starfish one-by-one and threw them back into the ocean. He repeated the behavior over and over again, despite the slew of starfish still left scattered across the beach. The boy called out, "There's too many; you are not making any difference!" When throwing a starfish back into the waves, the old man replied, "It made a difference for that one."

Incrementalism rarely stirs the heart. But, small improvements over time can amount to large reductions of effort, frustration, and abandonment. Not every problem needs to be solved to enhance a user's experience. Every bit helps. Now, that's wicked cool.

False Consensus Bias

For nearly 500 years, astronomers have embraced the belief that the Earth travels around the Sun. Nicolaus Copernicus wrote about his heliocentric theory in his 1543 book, *On the Revolutions of the Heavenly Spheres*.

Copernicus' work revolutionized astronomy. However, at the time, many people did not agree with his theories. Scholars and clergy ridiculed Copernicus, and in 1616, the Vatican banned his book. Luckily, Copernicus had the forethought to die on the day of its first publication. He avoided the worst his critics had to offer.

---

[8]Eiseley, Loren C. *The Star Thrower*. New York: Harcourt Brace & Company, 1979.

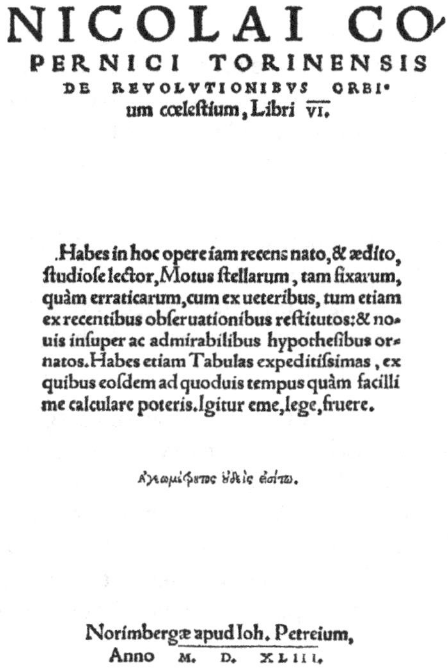

NICOLAI CO^

PERNICI TORINENSIS

DE REVOLVTIONIBVS ORBI•

um cœleftium, Libri VI.

.Habes in hoc opere iam recens nato, & ædito,
ftudiofe lector, Motus ftellarum, tam fixarum,
quàm erraticarum, cum ex ueteribus, tum etiam
ex recentibus obferuationibus reftitutos: & no-
uis infuper ac admirabilibus hypothefibus or-
natos. Habes etiam Tabulas expeditifsimas, ex
quibus eofdem ad quoduis tempus quàm facilli
me calculare poteris. Igitur eme, lege, fruere.

Ἀγεωμέτρητος ἠδεὶς ἐσίτω.

Norimbergæ apud Ioh. Petreium,
Anno   M.   D.   XLIII.

Figure 21-3. Title page of On the Revolutions of the Heavenly Spheres[9]

We can understand, and even partially excuse, the ignorance of the people in the 16th century. After all, they feared the unknown. Witches, lunar eclipses, and even red-haired people terrified the populations of Europe.[10] It is no wonder that the vastness of our universe frightened many of them as well.

Although witchcraft, celestial events, and gingers still scare some of us, we have advanced our thinking over time. We evolved. Yet, you may find it surprising that recent polls indicate that 20% of us still think the Earth is the center of the universe[11]. This data may seem erroneous, but also consider that it took until June 1999 for the Vatican to pardon Copernicus.[12]

---

[9]Cropped version of title page, Johannes Petreius 1543 edition of Nicolai Copernici torinensis De revolutionibus orbium coelestium, public domain.

[10]Institoris, Heinrich, and Jakob Sprenger. Malleus Maleficarum .. Francofvrti Ad Moenvm: N. Bassaeus, 1582.

[11]Dean, Cornelia. "Scientific Savvy? In U.S., Not Much." *The New York Times*. August 30, 2005. Accessed June 21, 2018. https://www.nytimes.com/2005/08/30/science/scientific-savvy-in-us-not-much.html.

[12]Ap. ""Heretical" Copernicus Reburied as a Hero." CBS News. May 22, 2010. Accessed June 08, 2018. https://www.cbsnews.com/news/heretical-copernicus-reburied-as-a-hero/.

We like to believe that others agree with us. In fact, the statement "we like to believe…" demonstrates this point. It is an overestimation based on my bias about what I think you and other readers might feel. Psychologists refer to this overestimation as the false consensus bias. We are bad at guessing how others feel and think. We regularly believe others share our political, social, and religious views. We even believe that all people know the Earth travels around the Sun.

We may never find ourselves debating the merits of heliocentric theory, but we will encounter false consensuses. Such bias happens frequently when designing software. We believe users share our knowledge. We expect them to behave like we do. We think users will act reasonably. Although these assumptions sound logical, we would be wrong.

False consensus biases compel us to test software designs. Users believe their views are reasonable and worldly, but they swim in a sea of biases. They lack empathy, believing software should address their specific needs, disregarding the various needs of other users. Without empathy, we design software in the same way. Neither the designer nor the user is immune to false consensuses. Empathy is our safeguard, because only when we validate our assumptions across a wide spectrum of other people can we determine if a belief is shared or individualized.

Fellow astronomers tested Copernicus' beliefs, and this gave us a greater understanding of our universe. Yet, it is important to remember that this understanding was not, and is not, universally shared. Even after 500 years, we still have a few holdouts. The future experiences you create will challenge your users' biases, your team's biases, and your own. Practice empathy and recognize that each of us sees the world from a unique vantage point. Sometimes these views align. Sometimes they do not. Empathy does not create consensus, but it does help keep our biases down to Earth.

Good Experience for All

The moral philosopher John Rawls spoke of empathy in his book *A Theory of Justice*.[13] Although he never directly referred to empathy in his text, he orchestrated a prime example of empathetic thinking. He constructed a thought experiment by asking people to design their own society.

---

[13]Rawls, John. *A Theory of Justice*. Cambridge, MA: Belknap Press of Harvard University Press, 1971.

In Rawls' experiment, you design your version of a perfect society. You must first decide how you want your society to function, choosing from among several possible freedoms, liberties, rules, regulations, and employment opportunities. For example, perhaps only women with a high IQ are permitted to vote. Maybe only citizens capable of 100 situps are allowed to eat fattening foods. Or, perhaps only heterosexual couples are entrusted to raise children. Once you are finished, you live in a society of your own making.

The catch: you have not been born yet and you do not know who you will be. You may be rich. You may be poor. You may be male or female, gay or straight, advantaged or disadvantaged. What type of society will you build to maximize your chances for happiness and fulfillment? Rawls contends that you would be best served to design a fair and just society, devoid of prejudice, intolerance, and bigotry. Because you never know—you could be a millionaire or a pauper, an Olympic athlete or lung transplant recipient, the next Albert Einstein or the next Homer Simpson. A perfect society is an empathetic one.

When we design an experience, we also look to perfect it. We create freedoms and rules, deciding what users can and cannot do. We allow users to read an article, or we erect a paywall. We permit users to opt out of receiving emails, or we spam them. We enable users to easily cancel their accounts, or we require them to run through a customer service gauntlet. Like Rawls' experiment, we would be best served to design a fair and just experience. If you were suddenly thrust into the user's role, would the experience be a good one? To craft a good experience for all users, we must empathize.

Some users use applications on behalf of other people. A parent typifies this type of secondary user. A child may be the primary audience for an educational app; however, a parent may configure the software, set parental controls, and chaperone the child's use of the app. The parent indirectly uses the software, as his or her experience is indirectly tied to the child's. When we minimize a secondary user's experience in favor of a primary audience, we alienate both audiences. Empathize with the needs of parents and you will enhance the connection with the child.

The salesperson-customer relationship is similar to that of a parent and child one. We need to empathize with a salesperson's needs just as much as a customer's. Tablet-based sales demos, estimate tools, product configuration apps, and guided purchasing all fall into this category of experiences. A salesperson may be a primary user when drafting an estimate, and become a secondary one when a customer reviews it. Likewise, the customer will be a secondary user until the salesperson relinquishes their control. The same happens when applying for a loan, buying a car, or visiting a doctor. If we wish secondary users to have a good experience, we must empathize with them, too.

In the marketplace, some users have one bad experience and never return. Industry estimates rank one-time app use as being as high as one in five. Apps become Chiclet-shaped gravestones, sitting idle and unnoticed, awaiting their inevitable deletion from the user's device. Designers of such experiences have failed to see their creations through another person's eyes.

We can never fully understand the needs of all users. They are too varied and numerous. Yet, when we design software, we still must account for a wide-ranging set of circumstances, aptitudes, and abilities. Our best tool for doing so is empathy.

When we empathize with users, we see their advantages and obstacles, their triumphs and struggles. And, if we are lucky, we sometimes get a glimpse into a better world.

Key Takeaways

- Empathy frames another person's experience as our own.

- Empathy requires us to understand another person's needs.

- Mirroring helps us understand another person's needs.

- Active listening is listening without judgment.

- We cannot empathize with those we do not hear.

- An answer to one wicked problem often creates another wicked problem.

- Through our repeated attempts to address the needs of particular users, we improve the experiences of all users.

- Empathy is our best tool to understand users' unique circumstances, aptitudes, and abilities.

Questions to Ask Yourself

- If I was suddenly thrust into the user's role, would my experience be a good one?

- How does my empathy for a person affect my perception of the person?

- Can I accurately mirror a person's body language and temperament?

- Am I listening to understand or to respond?

- Am I listening without judgment?

- Do my users have conflicting needs?

- Does solving a problem for one user cause a problem for another user?

- Am I trying to solve a wicked problem with reasoning alone?

- How can I improve the life of one user?

- Who are the secondary users (e.g., parents, support personnel, or salespeople) within an experience?

Authority

The 1963 Milgram experiment[1] at Yale University shocked the world. It showed how people readily yield to authority in the most frightening of ways. The experiment was simple. A participant sat at a table affixed with a microphone and a large control panel. The participant read aloud from a list of prepared questions. Behind a nearby screen sat another person who answered. If the respondent answered correctly, the experiment advanced to the next question. If the respondent answered incorrectly, the participant flipped one of the 30 switches on the control panel, thereby delivering a painful electric shock to the person sitting behind the curtain.

Over the course of several minutes, the participants were repeatedly instructed to increase the voltage. Each wrong answer caused another switch to be flipped. 15 volts. 30 volts. 45 volts. Higher and higher the voltage increased, as the person behind the screen wailed and screamed in agony. The participants squirmed in their chairs and pleaded to stop the experiment. Yet, they still followed orders. They still increased the voltage. In reality, the person wailing and screaming behind the wall was a paid actor and suffered no harm. Nevertheless, 65% of participants demonstrated that they would shock the actor to the point of unconsciousness. 450 volts. All it took was the instructor's demands. Abuses of authority can happen anywhere—in labs, war zones, offices, and even in user experiences.

---

[1]"PsycNET." American Psychological Association. Accessed June 08, 2018. http://psycnet.apa.org/record/1964-03472-001.

© Edward Stull 2018
E. Stull, *UX Fundamentals for Non-UX Professionals*,
https://doi.org/10.1007/978-1-4842-3811-0_22

On the other hand, authority can also help people make decisions, avoid dangers, and pursue goals. We trust a website is secure because of a certificate's authority. We believe a surgery is safe because of a doctor's authority. We hope driving directions are accurate because of a map's authority. Authority is integral to the functioning of everyday life. It either helps or hinders an experience, but we need not zap users to persuade them. We have other routes to persuasion.

Decision Fatigue

You sit at the piano. The lights shine upon you. You stretch out your arms and crack your knuckles. The conductor taps his baton on the podium. Tap. Tap. Tap. The audience quiets to silence. You take a deep breath, and the music begins. You play several notes from muscle memory, a few through conscious recall, and others by viewing the sheet music. You play and play, making decision after decision, going from chorus to bridge to melody and back. Every note creates an opportunity to be played off-key—to be played out of tune. The more you play, the greater the chance you will eventually fumble. Sweat lines your brow, you play and play again. Fatigue sets in. Your aching fingers strain to reach the farthest key and… Oh, no! Your hand slips and strikes the wrong note. The audience gasps. You stand back up and walk out of the room, never to return.

From playing a piano to inputting spreadsheet data, the quality of decisions decreases over time during an individual session.[2] The more decisions you make, the greater the likelihood you will eventually make a lousy one. And like a humbled concert pianist, a user may abandon her attempt to use a piece of software and never return to it again.

One of the more prevalent and mistaken beliefs about user experience is that users want interaction for the sake of interaction. This view often masks itself in the cloak of marketing—a phantom that runs onto the stage of a software interface, excites the audience, then retreats into the darkness. Unnecessary image carousels, overcrowded form fields, and futile sharing tools, are frequent culprits. Such distractions burden audiences with additional decisions to consider. Introduce too many considerations and you will fatigue your audience. They will make poor choices. They will blame themselves, as well as your software.

---

[2]Tierney, John. "Do You Suffer From Decision Fatigue?" *The New York Times*. August 17, 2011. Accessed June 21, 2018. https://www.nytimes.com/2011/08/21/magazine/do-you-suffer-from-decision-fatigue.html.

In earlier chapters, we discussed Hick's Law.[3] This law showed us that the number of choices within a given decision can sometimes work against the decision-making process. Though decision making is a complex endeavor, choices often slow users down. We witness Hick's Law at work when we observe users interacting with shopping carts: introduce a choice and you will create an opportunity for users to abandon their purchase (see Figure 22-1). Conversely, we see the benefit of reducing choices within the design of single-page websites. After all, if you want to avoid users getting lost within your site, one surefire strategy is to never take them anywhere.

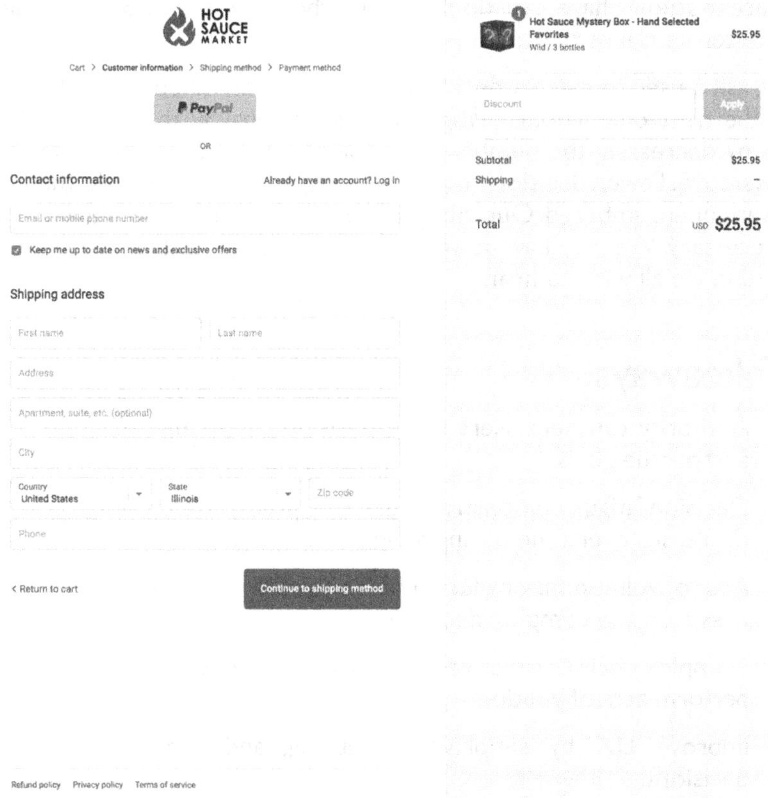

Figure 22-1. Shopping carts frequently display only content and navigation critical to check out and order completion[4]

---

[3]Soegaard, Mads. "Hick's Law: Making the Choice Easier for Users." The Interaction Design Foundation. February 2018. Accessed June 08, 2018. https://www.interaction-design.org/literature/article/hick-s-law-making-the-choice-easier-for-users.
[4]Customer Information Checkout. Digital image. Hot Sauce Market. Accessed June 7, 2018. https://hotsauce.market/.

Researchers have studied decision fatigue in relation to ego since the late 1990s. In a series of experiments[5] conducted by Dr. Roy Baumeister and his colleagues, researchers asked subjects to perform acts of willpower, such as resisting the temptation to eat chocolate cookies. Afterward, subjects were asked to solve complex puzzles. The experiment showed that after performing an act of volition (e.g., resisting eating a cookie), people were less able and willing to make complex decisions. Willpower and decision making appears to originate from a common pool of cognitive resources. Subsequent experiments also demonstrated the inverse relationship: complex decisions reduced subjects' ability to perform acts of volition. Decision fatigue and ego depletion go hand in hand. However, ego depletion is not without controversy. Several recent studies have cast doubt on the theory, whereas other studies further bolster its claims.[6]

When we ask a user to make a decision, we reduce her willpower and ability to make the next one. Though fatigue may seem unavoidable, we mitigate its effects by decreasing the number of decisions a user must make. Simplify. Reduce. Remove. Fewer decisions equal less fatigue. Our users glide through our software unencumbered. Our job is to design experiences containing only what is necessary. We may have fewer notes to play, but each will ring truer. That is a tune we all wish to hear.

Key Takeaways

- Authority can help users make decisions, avoid dangers, and pursue goals.

- Decision fatigue demonstrates that the quality of decisions decreases over time during an individual session.

- Acts of volition may reduce a person's ability and willingness to make complex decisions.

- Complex decisions may reduce a person's ability to make perform acts of volition.

- Improve UX by simplifying, reducing, and removing decisions.

---

[5]Baumeister, Roy F. "Where Has Your Willpower Gone?" *New Scientist* 213, no. 2849 (2012): 30-31. doi:10.1016/s0262-4079(12)60232-2.

[6]Engber, Daniel. "A Whole Field of Psychology Research May Be Bunk. Scientists Should Be Terrified." *Slate Magazine*, Slate, 6 Mar. 2016, www.slate.com/articles/health_and_science/cover_story/2016/03/ego_depletion_an_influential_theory_in_psychology_may_have_just_been_debunked.html.

Questions to Ask Yourself

- How can I simplify an experience at key decision points?
- When do important decisions occur within an experience?
- Are users forced to make decisions that occur in rapid succession?
- How can I stagger users' decision-making over the course of an experience?
- How can I stretch out the time between two decisions?
- Are users required to make too many decisions?

Motivation

In the late 1970s, Richard Petty and John Cacioppo studied how motivation affects persuasion. Their work resulted in their "elaboration likelihood model" (ELM).[1] We use the ELM in all sorts of decision making, from choosing a spouse to tapping a buy button.

Petty and Cacioppo proposed two primary routes to persuasion: a central and a peripheral route. When we are motivated to understand a subject, we are more likely to think about it and to elaborate on it. Petty and Cacioppo called this the central route. For example, say you are serving as a member of a jury, deciding the fate of a corrupt politician. Opposing lawyers offer their evidence. The FBI recorded the defendant demanding free spray tans from a New York salon. Empty bronzer bottles contain the defendant's fingerprints. Witnesses testify that they watched the defendant leave without paying. The defense attorney says his client was merely misunderstood. You weigh the testimony. You reread the transcripts, debate the merits, and deliver your verdict. Such careful consideration is the central route to persuasion.

People also have a peripheral route to follow. If the central route fails to persuade, people will consider surface characteristics. Perhaps you base your verdict on the attorney's expensive-looking suit, the judge's New Jersey accent, or the defendant's ridiculous comb over. These are cues. If a cue is present, a person may still be persuaded. The peripheral route is usually not as effective as the central route, but, over time, it still may persuade.

---

[1] Petty, Richard E., and John T. Cacioppo. "The Elaboration Likelihood Model of Persuasion." SpringerLink. January 01, 1986. Accessed June 08, 2018. https://link.springer.com/chapter/10.1007/978-1-4612-4964-1_1.

© Edward Stull 2018
E. Stull, *UX Fundamentals for Non-UX Professionals*,
https://doi.org/10.1007/978-1-4842-3811-0_23

Digital experiences leverage the central and peripheral routes. A website's detailed product information can provide a compelling case for making a purchase. Each fact serves as evidence—size, shape, weight, color, form, and function. A customer ruminates on this information, enhancing the product's persuasive effects along the central route. Attractive visual design, typography, and photography all contribute to an application's ability to persuade users, operating along the peripheral route (see Figure 23-1).

Figure 23-1. Facts on a product page persuade via the central route. The page's aesthetics persuade via the peripheral route[2]

---

[2]Hot Sauce Mystery Box - Hand Selected Favorites. Digital image. Hot Sauce Market. Accessed June 7, 2018. `https://hotsauce.market/collections/most-popular/ products/hot-sauce-mystery-box`.

Persuasion surrounds us. Thousands of messages compete for our attention, everything from billboard advertising to mobile app notifications. Each attempts to inform us of something new. However, depending on the route the message takes, its persuasive effects may quietly take hold over time or strike like a judge's gavel.

Key Takeaways

- Motivation affects persuasion.

- When people are motivated, they are more likely to think about a subject and to elaborate on it.

- When people are unmotivated, they are more likely to consider surface characteristics of a subject.

Questions to Ask Yourself

- What are my users' motivations?

- What evidence do users need to make a decision?

Relevancy

In the hills of Munnar, a subtropical highland town located at the southern tip of India, grows the Neelakurinji flower (see Figure 24-1). Nestled among lush tea plantations and shola grasslands, the plain looking shrub sways in the region's misty, temperate winds, rarely revealing its remarkably patient payload. The flower's secret lies hidden within its name: neela means blue in the Malayalam language. Every 12 years,[1] as if the flower were celebrating its last moments before becoming a teenager, Munnar's landscape explodes with light blue and purple Neelakurinji blossoms, transforming once-barren hillsides into electric blue confetti. The emergence marks a moment of time for the human beings living in the region as well, for the Muthuvan people use the flowering cycle to calculate their own ages.

---

[1]"Neelakurinji Blooming Season from August 2018 to October 2018 in Munnar." Kerala Tourism. Accessed June 08, 2018. https://www.keralatourism.org/neelakurinji/.

© Edward Stull 2018
E. Stull, *UX Fundamentals for Non-UX Professionals*,
https://doi.org/10.1007/978-1-4842-3811-0_24

Figure 24-1. A blossoming Neelakurinji flower[2]

A 12-year interval, give or take a few months, may seem to be an imprecise schedule when compared to the exacting timelines we use to create software. For many practical purposes, the cycle would be arbitrary and irrelevant. However, this interval works perfectly fine for observers of the Neelakurinji. Its relevancy is a matter of context. People expect the flower to bloom every 12 years, and, for them, the flower blooms at the exact right moment in time. Literary references to the flowering cycle date back to before the 17th century C.E. It may not be as fast as an app update, but the Neelakurinji has consistently hit its release deadlines for over 1,300 years.

Defining Relevancy

Terms like relevancy send epistemologists into a tizzy. What is relevancy, after all? It certainly sounds like something we would wish to achieve, but a precise definition is difficult to describe. One description of relevancy is that it is anything that furthers the completion of a goal. I like this one. If an experience helps a user complete her goal, it as relevant; if not, it is wasting her time.

---

[2]Photo by Aruna Radhakrishnan, "Neelakurinji (Strobilanthes Kunthiana)," October 5, 2008.

Relevancy = Time + Context

The relevancy of any particular item is dictated by both time and context. A hammer is relevant when you need to pound in a nail, but it is less so when you need to turn a screw. Likewise, features of an application are only relevant when you need them.

Feature Creep

Victorinox manufactures Swiss army knives. Since 1891, the knives have held the imaginations and the occasional salvations of explorers and would-be adventurers, alike. Astronauts flew with the knives into outer space. Submariners dove with them down to the ocean floor. Their ubiquity is only surpassed by their utility.

Victorinox's first knife contained four tools: a large blade, a reamer, a screwdriver, and a can opener. Today, the Wenger Giant Swiss Army Knife includes a telescopic pointer, flashlight, and compass, and 84 other tools.[3] Weighing in at over seven pounds, spanning nine inches, and costing $1,300, the knife is about as handy as a toothbrush nailed to a 2x4.

Although the Wenger Giant sells more as a collector's item than a functional knife, it does demonstrate how a useful and simple idea loses its usefulness over time through the addition of new features. I imagine that at some point over the years, someone thought, "hey, let's add a cigar cutter!" Following this epiphany, subsequent tools were added, ranging from chain rivet setters to club face cleaners. Gradually, the knife grew in size and weight; adding one feature after another reduced the knife's overall utility. For a knife is only as useful as its intended purpose: when you use its blade to open a delivery package, you are not using its corkscrew; when you use its flathead screwdriver to fix a toy, you are not using its can opener. Unneeded tools get in the way. Although a pocket knife containing five, or ten, or fifteen tools might be an acceptable tradeoff, 87 unique tools proves to be too much of a hindrance. Whereas the ratio of frequently useful to rarely useful tools is high with an original Swiss army knife, this ratio is lowered when more tools are added. As such, the extra weight and increased dimensions of the rarely useful tools become deadweight. Original Swiss army knives weigh a little under two ounces— equivalent to two AA batteries. In comparison, the two-pound Giant is equal to two, full-sized jars of mayonnaise. Which would you rather carry in your pocket all day?

---

[3]"Wenger 16999 Swiss Army Knife Giant - Most Expensive Item On Amazon." Amazon. Accessed June 22, 2018. https://www.amazon.com/Wenger-16999-Swiss-Knife-Giant/dp/B001DZTJRQ.

As software creators, we sometimes make the mistake of providing too many features. Screens become cluttered. Menus fill with options. Paragraphs burst along their seams. Irrelevant features not only clutter interfaces, but they also blind users to the content and functionality that they need.

In 2001, Apple offered Mac users a quick and easy way to rip, mix, and burn CDs of their favorite music. They called the application a digital jukebox and named it iTunes. Its brushed aluminum interface displayed only a handful of buttons, which came as no surprise because, as Steve Jobs described his competitors' offerings, "They are too complex, they are really difficult to learn and use.[4]" If only he had realized just how prescient his observation was about Apple's own creation.

In the years following, iTunes would add its Music Store, integrate with Windows, sell music videos, rent movies, carry eBooks, roll out iTunes University, recommend Genius picks, allow Home Sharing, launch Ping, wreck Ping, remove Ping, present iTunes Match, offer iTunes Radio, push Apple Music, and market the iCloud Music Library. Though robust, the application's user experience heaves under a thick blanket of bloated features, cryptic menus, and nonsensical signifiers. The once-promising elegance of a simple, scrolling list has been supplanted by competitive hierarchies and a fetishized affection for packaging art. iTunes contains a wealth of features; yet, the application struggles to make each feature relevant to a user's needs.

Like the tools of a Swiss army knife, software must balance the ratio of frequently and infrequently used features. We sometimes build too much, encumbering users with the unnecessary, forcing them to hunt and pick through an assortment of functionality not relevant to their goals. Everything a user does not need distracts her from everything she does.

Humor

Mark Twain once wrote, "Explaining humor is a lot like dissecting a frog, you learn a lot in the process, but in the end you kill it." If that statement is true, this chapter will be a bloodbath.

Humor is relevancy with a twist. We expect a particular outcome and—at the exact right moment—a twist delivers the unexpected. Uncertainty suddenly resolves, like a flower bursting into bloom.

---

[4]"Full Text of "Macworld March 2016"." Internet Archive. March 2006. Accessed June 22, 2018. https://archive.org/stream/Macworld_March_2016/Macworld_March_2016_djvu.txt.

Sigmund Freud believed the resulting pleasure of humor was the release of psychic energy.[5] Try to remember the last time you did not understand a joke, and you will recall just how satisfying it can be once you figured it out. Consider the wisecracking response the writer Dorothy Parker gave to a columnist when challenged to use horticulture in a sentence:

"You can lead a horticulture, but you can't make her think."[6]

The statement supplants one context for another. Horticulture leads us to an expectation: something involving plants or gardening. We certainly do not think of a person. Only when we read the well placed "her" do we revisit our previous context and compare it to our new understanding. "Her" transforms "hor"; "ti" becomes "to"; and now we find ourselves guiding a prostitute to culture. We derive pleasure through the rapid resolution of these incongruities.

The 18th-century philosopher Francis Hutcheson wrote of perceived incongruity as the basis for humor in his 1725 work *Thoughts on Laughter*. Although one person's humor frequently differs from another's, people may find humor at all times and in all places. People quit their jobs, divorce their spouses, and attend funerals. Yet, you would be hard-pressed to not find at least a glimmer of humor within these experiences. An employee quits a job and steals her favorite stapler. A spouse files for divorce and buys hair plugs. A mourner attends a funeral and his Tainted Love custom ring tone begins to play. Reconciling an old context with a new reality can take us into the dark corners of depression or the sunlight of humor. More often than not, we choose the sun.

Especially with technology, we have become attuned to intractable frustrations, such as the "PCLOAD Letter" printer error made famous in the 1999 movie *Office Space*. Michael Bolton, a character in the film, pummels a printer into the ground with a baseball bat to the score of Geto Boys' *Still*. Die muthafucka, die muthafucka, indeed. However, if we can intercept frustrating moments and turn them into humorous ones, we create relevancy with a distinctively human voice.

Versions of the Linux operating system add funny insults to error messaging, such as calling you a "Bonehead." YouTube may report "A team of highly trained monkeys have been dispatched" in response to server issues affecting the site. You may have even seen Tumblr's Tumblebeasts roam cartoon images of its datacenter. Each attempt at humor does not remedy all users' frustrations, but it does help them adapt to their new reality.

---

[5]Morreall, John. "Philosophy of Humor." Stanford Encyclopedia of Philosophy. November 20, 2012. Accessed June 22, 2018. https://plato.stanford.edu/entries/humor/.
[6]Keats, John. *You Might as Well Live: The Life and Times of Dorothy Parker*. Harmondsworth (Middlesex): Penguin Books, 1979.

Adaptation

Every few years, the Federal Highway Administration publishes the "Manual on Uniform Traffic Control Devices". Split into several sections, the 600-page guide (see Figure 24-2) describes the acceptable usage and design of American road signage. It is a fascinating read about safety, standards, and user experience.

Figure 24-2. Manual on Uniform Traffic Control Devices (MUTCD)[8]

Prior to the manual's publication in 1937, road signage ranged from colored ribbons to hand-painted posts. Though ubiquitous on today's roads, even the humble stop sign did not make an appearance until 1915,[7] which is remarkable considering that people had driven cars for several years before anyone thought that stopping them occasionally would be a good idea.

Signage excels at helping users adapt to new information. If it did not, we would have many more highway pileups. We can leverage some of the same techniques to improve the user experience of software.

---

[7]"History of the Stop Sign in America." Stop Sign History. Accessed June 22, 2018. http://signalfan.freeservers.com/road%20signs/stopsign.htm.
[8]Manual on Uniform Traffic Control Devices for Streets and Highways. Digital image. United States Department of Transportation Federal Highway Administration. Accessed June 7, 2018. https://mutcd.fhwa.dot.gov/mutcd_80_bday.htm.

Road signs communicate by form and content. Rectangles convey guidance. Pentagons signal schools. Circles indicate railroads. Triangles relay caution. Diamonds forewarn danger. Over time, these shapes inform our schemas for each type of message. We view the shape and it sets our expectation for the content contained within it. Just as the word "STOP" written within an octagon aligns with our expectation for a stop sign, so too does an underlined piece of text set our expectations for a website's hyperlink. People quickly understand meaning when form matches content, be it a highway sign or a sign-in button.

Reconsideration of a Goal

User experience is rarely a matter of going from Point A to B. Users confront obstacles, reconsider their goals, and sometimes redirect to new ones. Stores run out of stock. Auctions are outbid. Game characters die. Such events force users to reconsider their goals. They ask themselves: do I really need to buy this product… place this bid… complete this game? More often than not, they abandon the attempt. However, some users adapt.

Swiss psychologist and philosopher Jean Piaget first described how people adapt to new information.[9] His study of childhood development led to theories about how new information is either assimilated or accommodated. We assimilate information when it fits within our expectations. We accommodate information when it requires us to update or create new expectations.

Imagine you want to send flowers to a friend. You hop online, find a floral delivery website, and browse the available options. You choose a dozen chrysanthemums and click the buy button. A moment later, an alert pops up and tells you the flowers are out of stock. You are forced to create an entirely new expectation—you must accommodate to the website. To keep the user from abandoning her current experience, we need to find a detour: rather than accommodate, we want users to assimilate.

Think about the last time you encountered a detour while driving. Perhaps an exit was closed. Did you stop your car in the middle of the highway, step out, and walk away? No, you likely quickly noticed the road closure and took an alternate route. The alternative was more convenient than abandonment. You assimilated the new information and kept moving.

We do the same when using software. Instead of presenting the out-of-stock alert in our prior example, we could hide the chrysanthemums until they become available. Alternatively, we could suggest another product or a later delivery date. All of these approaches keep the user within her current experience, because the experience remains relevant to her needs.

---

[9]"Jean Piaget." Biography.com. February 19, 2016. Accessed June 08, 2018. https://www.biography.com/people/jean-piaget-9439915.

Relevancy redirects our behavior in curious ways. We may trek the flowered highlands of India, clobber a laser printer with a baseball bat, or zip down the highway looking for our exit. In every case, we pursue goals, seek relevancy, and adapt.

Key Takeaways

- Anything that helps users complete their goals is relevant.

- Features are only relevant when users need them.

- Everything a user does not need distracts her from everything she does.

- Use humor to alleviate frustrating user experiences.

- Users assimilate new information when it fits within their expectations.

- Users accommodate new information when it requires them to update or create new expectations.

- Users are more likely to abandon an experience when they accommodate new information.

Questions to Ask Yourself

- How can I make an experience more relevant to my users' needs?

- What might distract users from noticing vital information?

- Where can I introduce humor into a potentially frustrating situation (e.g., server downtime)?

- What are my users expectations for a product, service, or feature?

- What ways can I make an experience feel familiar to users?

- Which irrelevant features can I remove from an experience?

- Are my users assimilating or accommodating an experience?

Reciprocity

Years before the American President Richard Nixon stretched out his arms to form his famous V-sign, he was wrapping them around a pair of Chinese giant pandas named Ling-Ling and Hsing-Hsing[1] (see Figure 25-1). The two bears likely did not realize they were a part of a much larger embrace between two distant countries, brought together by international diplomacy and the power of reciprocation: a concept that spans borders, as well as every facet of user experience design.

Figure 25-1. A Chinese giant panda[2]

---

[1]"A Brief History of Giant Pandas at the Zoo." Smithsonian's National Zoo. February 21, 2017. Accessed June 08, 2018. https://nationalzoo.si.edu/animals/brief-history-giant-pandas-zoo.
[2]Skeeze. Panda Cub Wildlife. Digital image. Pixabay. February 27, 2015. Accessed June 7, 2018. https://pixabay.com/en/panda-cub-wildlife-zoo-cute-china-649938/.

© Edward Stull 2018
E. Stull, *UX Fundamentals for Non-UX Professionals,*
https://doi.org/10.1007/978-1-4842-3811-0_25

In 1972, Nixon visited China in an effort to normalize relations.[3] He and First Lady Pat Nixon, along with their diplomatic entourages, visited the Great Wall of China and major cities along the Yangtze River Delta. The pandas were a gift from then-leader of China, Mao Zedong. Ling-Ling and Hsing-Hsing were captured a few months prior in the far-off, temperate, central county of Baoxing, an area similar to the cool, misty forests of West Virginia, excepting the expansive fields of bamboo and the occasional black and white patched bear.

As soon as the two pandas arrived at the Smithsonian's National Zoo in mid-April 1972, they grabbed the hearts of the American public. Over 75,000 people visited the zoo the following weekend. Seen as a goodwill gesture by the Chinese government, the gift of the pandas was an overture harkening back through a thousand years of Chinese diplomatic history. The Chinese Empress Wu Zetian was said to have given pandas as a gift to the Japanese Emperor Tenmu[4] in the 7th century. Chinese diplomats revived this tradition in the 1950s and it continues today, primarily under loan agreements to national zoos and preservation programs.

Although historians refer to Nixon's trip and its gift as "panda diplomacy," it could just as easily have been referred to as "musk ox diplomacy," for the gift of the pandas was reciprocated by an equally furry gift to the Chinese. Nixon gave them Milton and Matilda, a pair of baby musk oxen. (Side note: you haven't seen adorable until you see musk oxen calves.) However, their trip was ill fated. Chinese zookeepers reported[5] the oxen arriving with runny noses, skin infections, and suffering from depression. After days in their new home, the musk oxen's distinctive shaggy fur began to fall out—the poor, little things. Depressed, runny-nosed, and now balding, sweet Milton and Matilda did not take up residence in the hearts of the Chinese public as much as their ursine peers had in America. Yet, this exchange shows us that even small gifts may create great opportunities that may last for decades, establishing relationships, reshaping economies, and setting in motion countless other experiences.

Two decades before Nixon's trip to China, a landmark French book had reached the shores of America. *The Gift* was written by the French sociologist Marcel Mauss. First translated into English in 1954, the book still stands as perhaps the most definitive explanation of our need to reciprocate. Mauss studied native societies. Although much of his work revolved around gift giving, his discoveries would eventually shed light on everything from marketing to software design.

---

[3]"Rapprochement with China, 1972." U.S. Department of State. Accessed June 08, 2018. https://history.state.gov/milestones/1969-1976/rapprochement-china.
[4]Beijing, Jon Watts in. "1,300 Years of Global Diplomacy Ends for China's Giant Pandas." *The Guardian*. September 14, 2007. Accessed June 08, 2018. https://www.theguardian.com/world/2007/sep/14/china.conservation.
[5]Besst, Nancy. *Milton and Matilda: The Musk Oxen Who Went to China*. San Francisco: China Books, 1982.

For his observations show us that gift giving creates bonds between givers and receivers. A gift creates an implicit obligation for the recipient to reciprocate. We give. We receive. We give back. Like intertwined threads, reciprocations weave into the fabric of an experience, strengthening and forming connections between givers and receivers, and by extension, between designers and users.

Marshall Sahlins' book, *Stone Age Economics*,[6] further expounded Mauss' notion of reciprocation by segmenting it into three types: generalized, balanced, and negative.

Generalized reciprocation is the most common: we give a gift and do not expect an immediate return. You might help a coworker with a project, shovel snow off your neighbor's driveway, or cook a meal for your spouse. Such gifts are freely given. We do not issue paper receipts expecting immediate repayment. It would be detrimental to our relationships if we did. "Dear spouse, I hope you enjoyed dinner. You now owe me one meal." Try it. You will soon realize this is a thread on which you do not wish to pull.

As working professionals, we can appreciate balanced reciprocation: we exchange our efforts for money. Tit for tat. Our employers or clients reciprocate the gift of our labors with the gift of a paycheck.

Negative reciprocation is expecting a gift without the intention of providing one in return. It is the equivalent of participating in your office's holiday gift-exchange but arriving empty-handed. "Happy holidays, I brought you nothing. Now, let's see what you have got for me." We will talk about how software often says the same in a few moments.

Generalized, balanced, and negative reciprocation may initially seem to operate only at a person-to-person level. However, they are a part of any interaction, including those between people and software. Mauss referred to the inherent quality of reciprocity in gift exchange as "total prestation." Granted, total prestation is not a term that easily slides off the tongue; yet, the term does speak to a gift being something greater than merely an object being given. We feel total prestation when we give and receive a gift. But what is it?

During Mauss' research, he encountered the gift-giving customs of the Māori people. Hau, the spirit of the gift, graced the both the giver and receiver. But, left unreciprocated, the hau would haunt the receiver like an evil spirit. It compelled the receiver to reciprocate. Although we may not feel haunted by the gifts we receive, we do feel obliged to reciprocate, be it person-to-person or person-to-software. Something is given, and we are compelled to return the favor.

---

[6]Sahlins, Marshall. *Stone Age Economics*. London: Routledge, 2004.

The Gift Exchange

When we design software, we create bonds between our users and ourselves. A user gives us the gift of their time and attention; we repay them with the gift of a good user experience. We give. They give back. Everyone reciprocates. These experiences contain generalized, balanced, and even negative reciprocations.

Some user experiences are generalized. As creators, we don't expect an immediate repayment by users. We lay the groundwork for future reciprocation, wishing to be repaid with more of the user's time and attention.

Other experiences are balanced. We give a user information or utility in direct exchange for performing a desired behavior, such as purchasing, sharing, or carrying out any number of other actions.

All too often, companies make the mistake of designing experiences with only negative reciprocations. Forced email sign-ups, unnecessary form fields, and takeover advertising are just a few of the ways companies extract a user's time and attention without reciprocating. Such experiences frustrate users. Time and attention are gifts that users may give us, but they should do only through a clear and balanced exchange. Negative reciprocation is a one-way street, paved with selfish intentions and littered with the wreckage of poorly designed user experiences.

Effective user experiences are comprised of both general and balanced exchanges. Consider the example of a free trial. If we follow the strict definition of reciprocation, we might believe our free trial is a balancedexchange between our application and the user: "Hello, user, here is a free trial. Now that you feel obliged, please buy our software." Tit for tat. But if we delve further into the concept of free trials, we soon realize this is a generalized reciprocation. A free trial establishes a possible future opportunity, but it takes time to mature.

By providing an incentive, users may reciprocate more quickly. You see this pattern play out frequently within consumer packaged goods promotions. A customer receives a free sample—say, for example, a tiny bottle of shampoo (a generalized reciprocation). The free sample contains a coupon (an incentive). The customer uses the coupon to buy a discounted, larger bottle of shampoo (a balanced reciprocation).

We can use a similar tactic to expedite a software purchase. A user receives a free 30-day software trial (a generalized reciprocation). The software could offer a discount to purchase a one-year software subscription (an incentive). The user then buys the discounted subscription (a balanced reciprocation).

Exchanges are often far subtler than a purchase. Enticing users to perform other behaviors follows the same forms of reciprocation. We lay the groundwork by providing engaging content and helpful functionality. For example, a calendar app may notify a user about upcoming events, seeming to only benefit the user. However, a user may choose to view details about the event, thereby directing his or her attention to a screen that also displays advertising. A video game may incentivize a player to "level-up" by providing the player with more powerful character traits, which in turn provides a reason to maintain a monthly subscription. Though subtle, each of these experiences is an exchange of gifts between the designer and the user.

The exchange of gifts between designers and users range from the exotic to the mundane. Like pandas plucked from a Chinese forest and deposited in front of curious onlookers, beautiful design and cutting-edge technology may fascinate and mesmerize users. We celebrate and reward such innovation. Yet, even the most unglamorous experiences can still reciprocate. The "Miltons and Matildas" of experiences repay the users' gifts of time and attention by providing unobstructed interfaces, clear communication, and a concern for the user's wellbeing. Such experiences serve as the basis for ongoing interaction.

Thus, we can understand reciprocation in UX to be less a matter of wowing users and more a means of creating a relationship with them. We give. They give back. The distance between designers and users shortens. That is the real gift.

Key Takeaways

- A gift creates an implicit obligation for the recipient to reciprocate.

- Incentives hasten reciprocation.

- An experience may include generalized, balanced, and negative reciprocations.

- Effective user experiences are comprised of both general and balanced exchanges between designers and users.

Questions to Ask Yourself

- What benefits do users derive from an experience?

- What benefits does a company derive from an experience?

- How can I incentivize users to pursue a goal?

- Does an experience fairly compensate users for their time and attentions?

Product

The ivory-billed woodpecker is one crazy-looking bird. A flaming crest of red feathers springs from its coal-black head. Two perfectly circular pale eyes affix to its face like a pair of tailored buttons (see Figure 26-1). Snowy trailing wing feathers offset its preened dark body. At nearly 20 inches long, the largest of its kind in the United States,[1] this woodpecker appears not so much to be a bird, but more a surprising flourish of stark, midcentury graphic design perched within the swampy virgin forests of the American South. Along with its dramatic appearance, the bird teaches us a crucial lesson about persuasion.

Figure 26-1. Artist's rendering of the ivory-billed woodpecker

---

[1]"Ivory-billed Woodpecker Identification, All About Birds, Cornell Lab of Ornithology." Photos and Videos, All About Birds, Cornell Lab of Ornithology. Accessed June 22, 2018. https://www.allaboutbirds.org/guide/Ivory-billed_Woodpecker/id.

© Edward Stull 2018
E. Stull, *UX Fundamentals for Non-UX Professionals*,
https://doi.org/10.1007/978-1-4842-3811-0_26

Currently believed to be extinct, the bird now commands the attention of ornithologists and birders from around the world.[2] In 2009, The Nature Conservancy, an international conservation organization, posted information about a $50,000 reward[3] for anyone who could verify the existence of a live ivory-billed woodpecker. Although incontrovertible evidence of the species was last recorded in 1938, finding a live specimen still captures people's hearts and imaginations. Several researchers have pursued the woodpecker for decades.

Universities, academic researchers, and impassioned amateurs heatedly debate the existence of the bird.[4] Naturalists claim to have recorded its signature sound, a loud double tap. Skeptics denounce the evidence. Counterarguments abound, but the controversial bird remains elusive.

An ivory-billed woodpecker symbolizes many things to many people: a reward, a pursuit, a cause. Where one person looks at the bird and only sees the $50,000 reward, another sees a lifelong pursuit, and yet another sees a once pristine wilderness overtaken by highways and suburban sprawl. Yet, all the while, the bird is just a bird. We make it something different: we transform it into an idea.

Idea Containers

You may think of products as being things such as toasters, iPhones, and Boeing 777s, but each is simply a container for an idea. We place our expectations, beliefs, and biases in these containers and give them a name, be it a bathtub or a button, a desktop application or a health club membership—or even an ivory-billed woodpecker. Just as a bird is just a bird, a product is just a product. We transform it into an idea.

Intrinsic and extrinsic cues shape our understanding of products. You see the clarity of an HD television screen, taste the sweetness of a Rainer cherry, smell the crisp scent of a new car's interior, feel the silky texture of a new sweater,

---

[2]Ploneadmin. "Cornell Lab of Ornithology." Elephant Evolution. March 28, 2016. Accessed June 08, 2018. http://www.birds.cornell.edu/ivory/.
[3]"Ivory-billed Woodpecker - The Search for the Ivory-Billed Woodpecker - Searching for the Ivory-Bill Bird - Ivory-bill Search | The Nature Conservancy." Red Foxes in Indiana | The Nature Conservancy. Accessed June 08, 2018. https://www.nature.org/ourinitiatives/regions/northamerica/unitedstates/arkansas/ivorybill/ivory-billed-woodpecker-the-search-for-the-ivory-billed-woodpecker.xml.
[4]Donahue, Michelle. "Possible Ivory-Billed Woodpecker Footage Breathes Life Into Extinction Debate." Audubon. January 25, 2017. Accessed June 08, 2018. https://www.audubon.org/news/possible-ivory-billed-woodpecker-footage-breathes-life-extinction-debate.

and hear the faint click of a keyboard's keystroke. We recognize extrinsic cues, too, such as a product's packaging and where the product is sold. We weigh all this evidence and issue a verdict: worthy or unworthy.

Applications face the same trials. An application is a product, but so are its parts. Tools and utilities within an application are products. Each was created with a purpose. Its purpose may be to provide an app's settings, a game's leaderboards, or a website's shopping cart. We evaluate each: "Is this thing worth my time?"

Users are selfish with their time. They will not read, visit, or interact with anything they perceive to be unworthy. People engage with only an idea of a product: this dress makes me pretty; this fruit juice improves my health; this app decreases my workload. But, if people think the idea is not worth their time, it never takes hold and the container remains empty.

Filling the Container

Theodore Levitt, an American economist and Harvard professor, popularized the saying "People don't want to buy a quarter-inch drill. They want a quarter-inch hole."[5] It is a witty aphorism about features and benefits. People use software because of the benefits it provides—not the features it offers. The benefits of an application are not always immediately apparent. You can stare at your computer, tablet, or phone all day and still have no idea if the applications contained within the device have benefits. Benefits of software are realized only through use.

So, how do we create products that will be used?

Now in its 15th edition, *Principles of Marketing*, a textbook written by Philip Kotler and Gary Armstrong, has taught countless students the fundamentals of marketing. Kotler provides a simple, but extensible, framework to understand any product: a product has a core product, an actual product, and an augmented product. (I take a few liberties to retrofit this framework for experience design.)

- A *core product* is the benefit of a product's use. Again, we will use a broad definition of the term "product" to mean a container for any idea. The core product of a weather app is the benefit of knowing how to dress today.

[5]Christensen, Clayton M., Scott Cook, and Taddy Hall. "What Customers Want from Your Products." HBS Working Knowledge. January 16, 2006. Accessed June 22, 2018. https://hbswk.hbs.edu/item/what-customers-want-from-your-products.

- An *actual product* is the product itself. The actual product of a weather app is the assortment of pixels and code that together comprise the application.

- An *augmented product* is the collection of related intangibles that add value to a product, such as customer service and technical support. The augmented product of a mobile app might include an online forum that addresses customer requests.

These descriptions of core, actual, and augmented products is marketing 101; however, these concepts have an immense impact on the creation of successful user experiences. As with many issues pertaining to marketing, we must separate the useful from the snake oil.

Applying the Framework

Imagine a heated debate taking place in your office. You and your team argue the merits of a website's contact form.

What is the core product of a contact form? (Remember: the core product is the product's benefit.) Is the core product the ability to enter contact information? Nope, not even close. Is the core product the ability to say something? Getting warmer. Is the core product the ability to receive a reply? Bingo! Users do not wish to supply their contact information arbitrarily. They expect an answer.

What is the actual product of a contact form? It is the design, copywriting, and underlying code.

What is the augmented product of a contact form? The website could automatically reply to the form submission with a brief thank-you email, containing an estimated time for a response. This email might show common FAQ questions related to the inquiry. Perhaps we ask the users to visit an online help forum. There are dozens of ways to augment a dull contact form.

Apply this framework of "Core, Actual, Augmented" to each piece of application functionality and watch it fill with ideas. Though this approach is "old-school" marketing, it is one of the more successful user experience design techniques available today.

Although creating a brilliant product is no small effort, it is only the first step. New products enter a marketplace saturated with good ideas. To soar above the rest, we must differentiate.

Differentiation

The German word "mittelstand" translates into English as "middle estate": not high, not low, but somewhere in the middle. It sits between the aristocracy and the lower classes. Mittelstand companies sit in the middle, as well: they are not multinational corporations, nor are they tiny shops, but they are highly specialized, small- and medium-sized companies. They form the backbone of the German economy, employing 60 percent of all workers.[6] Yet, despite these companies being collectively the largest segment of the economy, an individual mittelstand company is remarkably specialized, sometimes providing only one machine part to the global market—sometimes literally a cog in a wheel. They make one thing. Mittelstand companies differentiate themselves by making that one thing extremely well.

What can we learn from mittelstand companies? When we limit, we focus. When we focus on an experience, we are given the opportunity to design it extremely well.

Take, for example, the rather unsexy business of catching flies. The German company Aeroxon has been doing it for over 100 years. Theodor Kaiser, a German sweets manufacturer, founded the brand in 1909.[7] One can appreciate the need to catch flies in a candy factory. The company began with a single product: a fly catcher. They continually improved upon it, eventually creating a near-perfect fly catcher made of sticky taped, ribboned paper. By 1930, 130 million of these devices were being sold worldwide. They were not pretty, but they saved lives.

Before the widespread availability of penicillin, a fly catcher was your best defense against a host of afflictions, including diphtheria, typhoid, and cholera. You might look at the device as it hangs encrusted with flies on a sunny windowsill and say, "Ewww," but at least you would be alive. And in the century that has since passed, in the two world wars that followed, in the countless ups and downs in the economy that took place, Aeroxon still catches flies. They do one thing. They do it extremely well. It is the leading brand of household bug traps and insecticides in Germany.

The mittelstand approach works for software, too. Doing one thing well gives us focus. Through focus, we differentiate.

---

[6]Fear, Jeffrey. "The Secret behind Germany's Thriving 'Mittelstand' Businesses Is All in the Mindset." *The Conversation.* June 21, 2018. Accessed June 22, 2018. http://theconversation.com/the-secret-behind-germanys-thriving-mittelstand-businesses-is-all-in-the-mindset-25452.

[7]"History." Aeroxon Insect Control GmbH. Accessed June 08, 2018. https://www.aeroxon.de/en/company/history/.

Apple's App Store[8] and Google Play[9] now contain over two million apps each. Every year, developers release tens of thousands of other applications, ranging from Windows Phone to Blackberry, from desktop applications to embedded systems. You compete with everything, and that everything grows every day. Only the differentiated survive.

Consider the Twitter client app, Twitterrific by IconFactory. It battles a swarm of competitors, including Twitter's own free app. How does one compete with free? It differentiates. Although countless other Twitter apps have come and gone since Twitterrific's initial launch in 2007, the app maintains its position by providing a superior user experience. You would be hard-pressed to find easier ways to send replies and direct messages. Swipe right to reply (see Figure 26-2). Swipe left to message. Combined with an impressive array of customization features, the app flies high as one of the most-used—and arguably the most-preferred—Twitter apps available today.

Figure 26-2. Twitterrific's iOS app allows for quick replies and messages via swipe gestures[10]

---

[8]"App Store (iOS)." Wikipedia. June 08, 2018. Accessed June 08, 2018. https://en.wikipedia.org/wiki/App_Store_(iOS).
[9]"Google Play." Wikipedia. June 08, 2018. Accessed June 08, 2018. https://en.wikipedia.org/wiki/Google_Play.
[10]Twitterrific. Computer software. Version 5.19.2. Greensboro, NC: Icon Factory, 2018.

Doing one thing well is not limited to software creators alone, but to software users as well. Twitterrific's success may lie more in the hands of its users than its creators. The app enables people to tailor their experiences, allowing muting of hash tags, domains, and even phrases—goodbye, annoying memes! We can see similar reductions in all sorts of successful apps, from the distraction-free writing of Literature and Latte's Scrivener (see Figure 26-3) to the sparse lists of Realmac's Clear (see Figure 26-4).

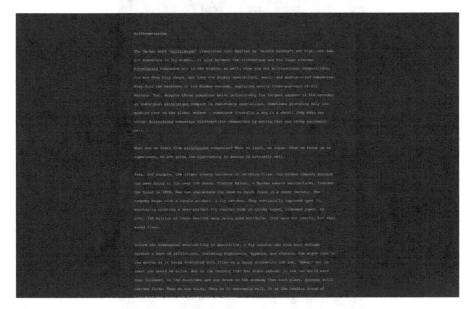

Figure 26-3. The distraction-free, full screen composition mode of Literature & Latte's Scrivener 3[11]

---

[11]Scrivener 3. Version 3.0.2. Cornwall, UK: Literature & Latte, 2018.

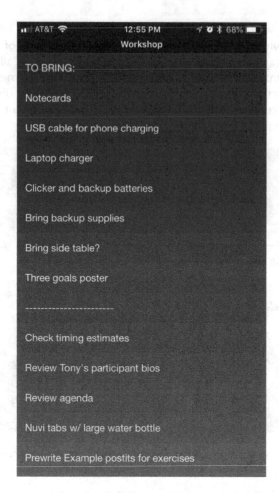

Figure 26-4. The elegant and sparse visual design of Realmac's Clear differentiates the app against a slew of competitors[12]

Like a mittelstand company, whatever experience you create—one or a thousand—you need to design each extremely well. For any experience can distract if not handled with care. Focus on what users truly need and remove all else. Doing so distinguishes your product from your competition. Do anything short of that and you are simply swatting at flies.

---

[12]Clear. Version 1.7.6. Brighton, England: Realmac, 2017.

Key Takeaways

- A product is merely an idea container.

- Intrinsic and extrinsic cues shape our understanding of products.

- Users will not read, visit, or interact with anything they perceive to be unworthy of their time.

- A core product is the benefit of a product's use.

- An actual product is the product itself.

- An augmented product is the collection of related intangibles that add value to a product, such as customer service.

- Great products do one thing extremely well.

Questions to Ask Yourself

- What are the product's intrinsic and extrinsic cues?

- What benefits do people derive from using the product?

- What are the product's benefits not related to use (e.g., great customer service)?

- What one thing does the product do better than its competitors?

Price

In Leo Tolstoy's short story, *Ivan the Fool*, three brothers seek success in three different ways: Simon wages war, Tarás pursues riches, and Ivan works the land. But like all experiences, each endeavor extracts a price.

Simon and Tarás wish to leave their family's estate to pursue conquest and fortune. They bully Ivan, demanding money and food from him to support their adventures. Ivan is simple, having little to do with his brothers' interests, but he tells them to take what they want.

Shortly after Simon and Tarás leave home, the Devil begins to create strife among the three brothers. He pits one brother against the other, plaguing the men with losses and misfortunes. Simon trades his honor for prestige. His hubris leads him to military defeat. Tarás trades his integrity for gold. His misdeeds lead him to financial ruin. However, the Devil is unable to make Ivan fail.

The Devil twists Ivan's plow and floods his fields, but Ivan persists, facing each challenge with kindness and humility. Ivan's good nature and simplistic views protect him from the Devil, ultimately leading Ivan to rule his own kingdom. The kingdom forsakes gold and glory, but in return, Ivan and his subjects live in harmony. Not a bad conclusion for a Russian fairytale.

Tolstoy's allegory demonstrates that the price of something is more than what we literally pay. It is an exchange of values: give and take. Time, energy, attention, and money are just a few of the many currencies we can use. What we are willing to exchange defines an experience, be it a single interaction or a lifetime. Each exchange comes at a cost: we save time by sacrificing quality; we gain convenience by decreasing privacy; we build communities by surrendering authority. Nothing is free.

© Edward Stull 2018
E. Stull, *UX Fundamentals for Non-UX Professionals*,
https://doi.org/10.1007/978-1-4842-3811-0_27

Although Tolstoy wrote *Ivan the Fool* in 1885, we can still see modern day organizations fulfilling each of the three brotherly roles. Some companies act like military generals, believing they can simply command people to use their products, discarding their users' goals in favor of their own business' objectives. Such companies fail quickly. Other companies play the part of rapacious swindlers, focusing solely on short-term financial gains, squandering their user's time and trust with unethical tricks and dark patterns. Such companies fail eventually. Yet, we realize the greatest success by being the virtuous fool, taking no experience for granted, working on behalf of our users, and serving our audiences with kindness and humility.

Kindness and humility begin our discussion on price. For price, at its core, is the measure of any relationship, comprising both give and take, both gains and losses, both benefits and costs. Price defines a relationship, and experience becomes our ledger.

Economy of Needs

At the turn of the 20th century, an Italian economist named Vilfredo Pareto created a power law that we still use today. A power law bases itself on two quantities: one fixed, one proportional. You experience a power law each morning if you are a coffee drinker. Some mornings you may drink from a small cup and use one teaspoon of sugar. Other mornings you drink from a large mug and add several teaspoons. The more coffee you pour, the more sugar is needed. The quantity of one dictates the quantity of another. Pareto's power law involved the relationship between population and land ownership. He recognized that 20% of Italy's population owned 80% of Italy's land. These observations later extended into other studies, such as the relationship between income and taxes. In each case, Pareto saw that approximately 20% of causes generated 80% of the effects (see Figure 27-1).

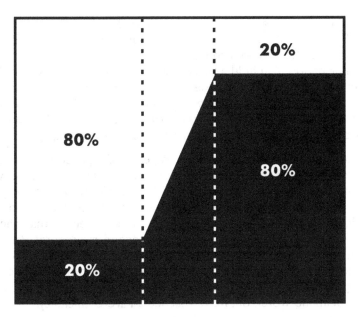

Figure 27-1. The 80/20 rule, where approximately 20% of causes generate 80% of effects

In 1941, Joseph Juran[1] formed his theory of "the vital few and the trivial many," based on Pareto's distributions. Commonly known today as the Pareto principle, or the 80/20 rule, we see these distributions in everything from the relationship between top salespeople and company revenue, to the relationship between earthquake damage and seismic activity. We see examples of the Pareto principle everywhere. Twenty percent of products generate 80% of a company's revenue. We wear 20% of our wardrobe 80% of the time. You'll likely find 20% of this book delivers 80% of its value.

Pareto distributions are not always 80/20. However, they always feature an unequal weighting where a short peak of high values and a long slope of low values are present. Imagine a mountain cut in half. The mountain may be tall and thin, or short and wide. Every mountain has a peak, and every mountain has a slope. Now, consider a digital project. The short peak of high values reflects your vital efforts; the long slope reflects all the extraneous— and sometimes trivial—activities. We can almost feel Pareto distributions at play in our project schedules: we spend long portions of time caught up in minutiae (80%), and we cram the most important work into the short time before a deadline (20%). Consequently, that 20% period of time leads to 80% of our most important work. We can change this, however. Pareto isn't predestination.

[1]"Joseph Juran." *The Economist.* June 19, 2009. Accessed June 08, 2018. https://www. economist.com/node/13881008.

Pareto is also known—at least by economists—for a concept called Pareto efficiency. Pareto efficiency involves resource allocation. Picture two classrooms full of children. Each classroom contains rows of desks and smiling kids waiting in eager anticipation. You hold a basket full of cookies. You walk into the first classroom and hand a kid a cookie. Exiting the room, you walk down the hallway and into the second classroom. There, too, you hand a kid a cookie and exit the room. Simple process: walk into a classroom, hand a kid a cookie, and walk out. You repeat the process until your basket is empty.

With 10 cookies, the math is simple: you repeated the process five times. The fifth time is said to be Pareto efficient—you gave away all your cookies with an equal allocation. But what if you started with 11 cookies? On the sixth repeat of the process, one classroom gets a cookie, and the second classroom gets nothing. It isn't Pareto efficient. One classroom is better off than the other. How can we improve this situation for the second classroom? Perhaps we could voluntarily compensate the second classroom somehow, such as by given the kids extra recess time. If we did, this compensation would be called a Pareto improvement. With the second classroom now better off and the first classroom being no worse off, we've reached a Pareto efficiency.

Although Pareto efficiencies drive free markets, they also create an argument for fairness: we wish all parties in an exchange to be made whole—no party should be made better off at the expense of the other.

Pareto UX

In terms of user experience, you could say that Pareto efficiency involves the fulfillment of needs—for users as well as businesses. We wish for both users and businesses to be made whole. Neither is better off at the expense of the other. Needs between users and businesses occasionally conflict. What a user wants, a business may not, and vice versa. Imagine an app that performs one of two functions: a user presses a button and gives $1 to himself, or a user presses a button and gives $1 to the business. Obviously, a user would prefer the one outcome and a business the other. The app is not Pareto efficient. When one party is better off, the other is worse off. We need to find an improvement to achieve the efficiency.

Like the "Acme $1 app," every application provides a value exchange between a user and a business. Sometimes this exchange is front-loaded, and users pay for the application before its use. But often this exchange is less direct. A business may offer utility in exchange for a user's loyalty. It may offer entertainment in exchange for attention. It may offer content in exchange for information. In each case, the Pareto efficiency illuminates the exchange: does the exchange benefit one party over the other? Is it optimal? Is it fair?

Complex applications may contain dozens—if not hundreds—of value exchanges between a user and a business. A visitor to an e-commerce site is offered utility in exchange for attention, information, and loyalty. The exchange is worthwhile for the business, if the cost of providing the utility is matched by the value it receives from users. However, these exchanges are easily mishandled when a business asks for too much from its users. Many users are a form field away from abandonment. Attention, too, is frequently overtaxed by obnoxious marketing. Moreover, the exchange of personal information threatens to surpass any value offered by a business.

If an application benefits the business or user at the expense of the other, we must revisit Pareto's concept of efficiency and find improvements. How do we know if there is an unbalance? We create a list.

1. Sale of products	(+Business, +Users)
2. Customer information	(+Business, -Users)
3. Brand exposure	(+Business, +Users)
4. Ad impressions	(+Business, -Users)
5. Discounts	(-Business, +Users)
6. Free shipping	(-Business, +Users)
[...]	

Using this example, we can see that the sale of products benefits both the business and user. A business gets paid; the user gets the product. This exchange is efficient. However, when we evaluate the exchange of customer information, the business benefits and the user does not. The user gave up a bit of their privacy—but in exchange for what? How do we find an improvement to make this exchange efficient?

Unequal Exchanges

In cases where an unequal exchange is made, we need to find a means of compensation. When a user gives up a bit of privacy, we can compensate her. Perhaps her private information is necessary to complete an order. The business profits from this information. But we could use it to expedite forms for the user, such as auto-filling a shipping address. Such functionality now provides benefits to both the business and user. The key is to compensate for every unequal exchange.

You can scoot by with one or two unequal exchanges, but push it too far and people will abandon. A good example is advertising placements. Ad impressions clearly benefit a business through increased awareness and possible ad network payments, but even the most jaded marketer must admit that advertising rarely benefits a user. Where can Pareto improvements be made in this equation? The short answer is that sometimes they cannot. To offset the cost of some experiences, we must overcompensate in other areas. Google Search does a fine job of this. Advertising surrounds it users, but the resulting utility is so great that the exchange is welcomed by those who carry out the 3.5 billion searches per day.[2]

At first, Pareto efficiencies may seem unnecessarily analytical, but they can help you reveal future user experience issues before they occur. Strive for balanced exchanges, for an imbalanced one will always remedy itself eventually: you'll either fail by offering too much, or fail by offering too little. Balance maintains fairness. In the economy of needs, fairness always wins.

Contrasts and Anchors

I have a question for you: is the oldest performing ballerina more than 25 years old? If so, by how much? Without cheating, please take a guess now. I will wait.

Waiting…

What did you guess? Maybe you added 10 years; a ballerina at 35 seems reasonable. Maybe you added 20 years; a ballerina at 45 seems possible. Would you have added 61 years? Likely not. However, the oldest performing ballerina is 86-year-old Grete Brunvoll from Norway.[3]

Although regularly performing ballet at age 86 is remarkable, your answer skewed lower because of the anchoring effect the question: "Is the oldest performing ballerina more than 25 years old?" Even though 25 years old perhaps seemed too young to retire, the number still anchored your expectations.

Anchoring affects everything from national budget policies to the price of tap shoes. It is a well-known tactic in the restaurant industry. Ever wonder why a $99 bottle of champagne is featured next to $5 chicken wings on a sports bar's menu? The high anchor price of the champagne makes everything look less expensive. The inverse is true, as well: feature a low-cost item, and even the most reasonably priced items will appear expensive.

---

[2]"Google Searches Per Month." Google. Accessed June 22, 2018. https://www.thinkwithgoogle.com/data-gallery/detail/google-searches-per-month/.
[3]"Oldest Performing Ballerina." Guinness World Records. Accessed June 08, 2018. http://www.guinnessworldrecords.com/world-records/oldest-performing-ballerina.

The human brain acts as a biological cash register, recording the highs and lows—the costs and benefits—of commercial experiences. Psychologists and professors, such as Robert Cialdini, Paco Underhill, and G. Richard Shell, have dedicated their careers to understanding influence and shopping behavior.

Behavioral economists and psychologists, such as Daniel Kahneman, Amos Tversky, and Richard Thaler, use pricing to delineate the very fabric of human behavior and decision making, proving price is more than a monetary measurement. We measure prices in dollars, but also time, attention, pleasure, and risk. As such, contrasts and anchors affect how we perceive costs.

Decoy Effects

Decoy effects steer people to choose one of two options by presenting a less preferable third option. The third option creates a decoy. Such decoys offset all sorts of questions, from software subscriptions (e.g., $1, $10, $30—decoy) to video game weapons (e.g., epic, uncommon, poor-quality—decoy).

Contrast and anchoring may trick users into making unwise decisions, but we can use the same techniques to create a greater good.

Ethical Anchoring

Attaching anchors to data creates a bias. But biases are not inherently bad. We can use ethical anchoring to persuade users to perform beneficial tasks.

Rather than display a progress bar at 0% completion, we can begin it 20% completed. We give users a head start. The partial completion creates an anchor that persuades users to keep going. After all, the first step in a journey is often the hardest one. Skip this step, and your users will be well on their way to a goal.

Want users to share your content? Provide them a pre-filled message (see Figure 27-2). Not only does pre-filling save users time, but it also gives users an example of the type of message that could be shared. Users add a personal flourish, edit, or delete. Whatever the decision, a pre-filled message sets an expectation of how something works.

Figure 27-2. Pre-populated tweet content created through publish.twitter.com[4]

Setting an expectation helps users visualize the future, be it a download, a tweet, or a dance recital. Contrasts and anchors inform decisions by supplying information where none exists. When we anchor information, we reduce a user's need to figure things out for themselves, quickening the journey to reach their goals, turning pirouettes into promenades.

Highly Destructive Operations

Years before the actor Sir Alec Guinness donned a cloak and wielded a light saber, he stood on the embankment of Sri Lanka's Kelani River and shouted instructions to his fellow British prisoners of war. A largely fictionalized retelling of the construction of the Burma–Siam railway, *The Bridge on the River Kwai*[5] recounted the hardships, struggles, and ultimate triumphs of Allied POWs and conscripts under the brutal occupying forces of Imperial Japan during WWII. Sir Alec was its star. The film shines as an example of both the foibles of pride and the exultations of sacrifice, for it tells the story of how the Allied prisoners painstakingly crafted an Axis rail bridge, only to destroy the same bridge once it was completed. It is a vital lesson shared by all creative endeavors: everything built is eventually destroyed—sometimes by others, other times by ourselves.

---

[4]"Guides - Twitter Developers." Twitter. Accessed June 10, 2018. https://developer.twitter.com/en/docs/twitter-for-websites/tweet-button/overview.
[5]*The Bridge on the River Kwai.* Directed by David Lean. Produced by Sam Spiegel. By Pierre Boulle. Performed by William Holden, Alec Guinness, Jack Hawkins, and Sesshu Hayakawa. United States: Columbia Pictures Corp., 1957.

Users build and destroy. They spend hours finding the perfect product, only to abandon it in a shopping cart. They spend minutes filling out a lengthy form, only to cancel it midway. They spend seconds viewing an app's login, only to ignore it and never return again. Users destroy, burn, and disassemble their own experiences repeatedly. They throw away their time, energy, and attention. They build and discard. They voluntarily dynamite the bridge leading to their own goals.

In the movie, Sir Alec's character, Colonel Nicholson, experiences a moment of revelation when he realizes all his efforts have been wasted in the pursuit of an unworthy goal. In his last breath before dying, he exclaims, "What have I done?" falling onto an explosive plunger, detonating the bridge he had just completed. Though he had labored for months, he abandoned his goal within seconds of his epiphany.

Sudden abandonment is an ever-present liability within an application. It is the bomb hidden within every user experience. We construct experiences, bridging user needs with user goals. Yet, users alone decide whether to continue their journeys or to abandon them—they either traverse the bridge or blow it up.

Consider the fate of Myspace. At its height in 2006, the social site attracted more daily users than Google.[6] Over 75 million people visited the site each month, roughly equivalent to two years of traffic on the Golden Gate Bridge. Since then, the number of Myspace accounts has dwindled. Think of the millions of users sitting in front of their computer screens, reflecting on all the time they had spent on Myspace, and in a moment of revelation exclaiming "What have I done?" then tapping the "Delete Account" button. Purchased for $580 million in 2005, News Corp sold Myspace for $35 million in 2013.[7] Kaboom.

While the shift from one social network to another is certainly not new, it does highlight highly destructive operations. Even when users spend considerable time and energy investing in an experience, they will still abandon it. As the saying goes, "It's a matter of when, not if." Facebook gained when Myspace lost. Myspace gained when Friendster lost. Friendster gained users from now defunct networks such as PlanetAll, Bolt, and SixDegrees. Everyone eventually abandons.

E-commerce fares no better. Companies have agonized over shopping cart abandonments since the birth of online shopping. Several studies estimate abandonment affects two-thirds of all online transactions. According to 2014

---

[6]Arrington, Michael. "MySpace, The 27.4 Billion Pound Gorilla." TechCrunch. June 13, 2006. Accessed June 08, 2018. https://techcrunch.com/2006/06/13/myspace-the-27-billion-pound-gorilla/.

[7]Vascellaro, Jessica E., Emily Steel, and Russell Adams. "News Corp. Sells Myspace for a Song." *The Wall Street Journal*. June 30, 2011. Accessed June 08, 2018. https://www.wsj.com/articles/SB10001424052702304584004576415932273770852.

U.S. Census data, e-commerce totaled 394 billion dollars in retail sales;[8] yet these dollars only account for one third of all carts.[9] Without abandons, U.S. online sales would top 1.18 trillion—slightly more than the total GDP of Mexico.

Everyone abandons, but how do we preserve relationships with users in the interim? How does each interaction with a user contribute to or take away from the overall experience?

In the parlance of marketing, we define the preservation of a relationship as loyalty. A huge category in its own right, loyalty also plays a vital role within service and user experience design.

Avoid Mistrust

Security and privacy breaches affect relationships with users. When hackers stole data for an estimated 40 million card accounts and 70 million records of guest information from Target[10] during the holiday season of 2013, the company reeled from plummeting customer satisfaction scores, stock price drops, a downgrading of its credit rating, and the eventual replacement of its CEO. However, the full impact of this breach may continue to reverberate for years. How willing are you to use Target's website: enough to apply for a store card? Enough to create a bridal registry? Enough to buy?

The 2016 American presidential election flooded airwaves with a slurry of gossip, misinformation, and conspiracy theories. From PizzaGate to hacked emails, voters were inundated. As voters were often also Facebook users, inundation came in the form of eye-raising posts and hair-pulling comments threads. Our trust eroded. What we relied on as a daily diversion nearly capsized a democracy. Likes became contentious endorsements, worthy of heated debates and creative name-calling. Goofy personality quizzes,[11] weaponized by campaigns, became sophisticated social engineering tools

---

[8]*QUARTERLY RETAIL E-COMMERCE SALES 4TH QUARTER 2014*. PDF. Washington, D.C.: The Census Bureau of the Department of Commerce, February 17, 2015. https://www2. census.gov/retail/releases/historical/ecomm/14q4.pdf

[9]"Why Online Retailers Are Losing 67.45% of Sales and What to Do About It." Shopify Content-ForAmbitiousPeoplelikeYouAugust6,2013AccessedJune22,2018.https://www.shopify.com/ blog/8484093-why-online-retailers-are-losing-67-45-of-sales-and-what-to-do-about-it.

[10]Yang, Jia Lynn, and Amrita Jayakumar. "Target Says up to 70 Million More Customers Were Hit by December Data Breach." The Washington Post. January 10, 2014. Accessed June 08, 2018. https://www.washingtonpost.com/business/economy/target-says-70-million-customers-were-hit-by-dec-data-breach-more-than-first-reported/2014/01/10/0ada1026-79fe-11e3-8963-b4b654bcc9b2_story. html?utm_term=.5c6a6d4e11c1.

[11]Brodwin, Erin. "Here's the Personality Test Cambridge Analytica Had Facebook Users Take." *Business Insider*. March 19, 2018. Accessed June 22, 2018. http://www.businessinsider.com/ facebook-personality-test-cambridge-analytica-data-trump-election-2018-3.

capable of piquing interests and exacerbating divisions. Whistleblowers would come to reveal the extent of this manipulation. Facebook's audit indicated 87 million affected accounts.[12] Despite the company's eventual highly publicized apologies and congressional testimony, its problem with mistrust began more than a decade earlier.

In late 2007, Facebook launched the advertising platform Beacon. Beacon worked with partner companies, such as Blockbuster and Overstock.com, to extract knowledge of website visitors' activities—including buying and renting products and signing up for accounts. This detailed information was subsequently broadcasted in the newsfeeds of other Facebook users. Surprise gift purchases and movie rentals began to show up in the newsfeeds of wide-eyed spouses and voyeuristic friends across the social network. Years later, Facebook would terminate the service as part of a legal settlement, along with a 9.5-million-dollar judgment[13]—a drop in the bucket for a company with a market cap approaching 405 billion dollars in 2016. Yet, we must ask ourselves how such negative experiences change a brand. Some users love Facebook, some hate it, and many mistrust it. You can almost hear the faint sound of the Kwai River flowing beneath many Facebook user experiences.

Mistrust manifests in subtler ways as well. Where persuasion illuminates a path, manipulation dims it. Manipulation diverts users into making inadvisable decisions. We see its hand in ads placed near buttons, as unscrupulous designers attempt to capture a users' mis-clicks. We see manipulation in link-bait titles, such as "The most important issue you must deal with today!", "10 things no one will tell you", and "You won't believe this actually exists!" Hyperbole may provide a momentary spike in traffic, but it will ultimately erode your user's trust. Fool me once, shame on you; fool me twice, shame on me; fool me three times, I'll never click one of your damn links again.

Receive and Respond

Before online surveys, before market analysis, before focus groups, before telephone polling, before interviews, and before any research at all, was conversation. Conversation forms the heart of all relationships between human beings. From a loud "I love you" to a silent shuffle in your chair, we exchange information with one another by conversing.

---

[12]Salinas, Sara. "Facebook Says the Number of Users Affected by Cambridge Analytica Data Leak Is 87 Million." CNBC. April 04, 2018. Accessed June 08, 2018. https://www.cnbc.com/2018/04/04/facebook-updates-the-number-of-users-impacted-by-cambridge-analytica-leak-to-87-million-.html.
[13]Kravets, David. "Facebook's $9.5 Million 'Beacon' Settlement Approved." *Wired.* June 03, 2017. Accessed June 08, 2018. https://www.wired.com/2012/09/beacon-settlement-approved/.

We sometimes find ourselves so engaged in a conversation that time slips by like a fast-flowing stream of consciousness. Conversely, some conversations drip slowly, like a leaky faucet. Mihaly Csikszentmihalyi described in his book, *Flow: The Psychology of Optimal Experience*, that a person's perception of time is altered by her or his focused attention. When we immerse ourselves in a pursuit, we maintain a state of flow.

Software enhances flow when users receive immediate feedback throughout an experience. Applications receive all sorts of inputs: mouse clicks, gestures, form field blurs, interval timers, and the like. However, these same applications often fail to respond with any sort of tangible feedback to their users. Akin to a user and an application passing each other in the hallway, the user says, "Hello, application!" and the application walks by without giving the user the slightest acknowledgment, making an application's personality appear cold and mechanical.

Dan Saffer's book, *Microinteractions: Designing with Details*, refers to software feedback as a "personality-delivery mechanism." It affords us the ability to interject experiences with a human presence, including a full range of dispositions like humor, warmth, and charisma. Applications receive digital inputs, but they respond to human beings. It would serve us well to remember this. After all, users abandon for two reasons: when they dislike what they experience, and when they experience nothing at all.

A destructive experience extracts a cost from both users and creators. We pay the price, spending time, energy, and attention. Occasionally, we must clear the way for new experiences by destroying the old; yet, we must still maintain our connections with users. Trust and communication keeps the bridges intact.

Key Takeaways

- Good UX takes no experience for granted, serving its audiences with kindness and humility.

- Complex applications may contain hundreds of value exchanges between a user and a business.

- To offset the cost of some experiences, we must overcompensate in other areas.

- Imbalanced exchanges cause products to fail by offering users too much or too little.

- Anchoring affects users' perceptions of prices.

- Ethical anchoring persuades users to perform beneficial tasks.

- Users abandon experiences, even when users invest considerable time and money in an experience.

- Users abandon for two reasons: when they dislike what they experience, and when they experience nothing at all.

Questions to Ask Yourself

- Which 20% of an experience provides the most value to users?

- Where within an experience do users spend 80% of their time?

- Is an experience Pareto efficient?

- How can I employ ethical anchoring techniques to help users pursue their goals?

- Where within an experience is a user likely to abandon?

- How does each interaction with a user affect her or his overall experience?

- Does an experience persuade users or manipulate them?

- Could users interpret any part of the experience as deceitful or unsafe?

- How can I help users maintain a state of flow?

- Ultimately, is the experience worth its price?

Promotion

Since before the Bronze Age, East Asian cultures have cultivated rice. They have grown, harvested, stored, and eaten it, written on it, sung about it, and painted it. Therefore, we should not be surprised to learn that people have spent a great deal of their time perfecting rice farming (see Figure 28-1). Such close attention had been paid to its production and cultivation that rice became an early metaphor for many aspects of living, including life itself. It even illustrates promotion.

Figure 28-1. A terraced rice farm in Japan[1]

[1]DeltaWorks. Japan Rice Terraces Kumamato. Digital image. Pixabay. June 15, 2015. Accessed June 7, 2018. https://pixabay.com/en/japan-rice-terraces-kumamoto-green-808990/.

© Edward Stull 2018
E. Stull, *UX Fundamentals for Non-UX Professionals*,
https://doi.org/10.1007/978-1-4842-3811-0_28

Mencius, a Chinese philosopher from the 4th century BCE, described human nature in his poem *Pulling up Sprouts*.[2] Mencius' teachings dealt with two competing perspectives about the goodness of people. Taoists champion the first perspective: you sow a seed then leave it to grow (ignore). They believe humans need only to realize their innate goodness to become good. Confucians defend the second perspective: you sow a seed then cultivate the plant with excruciating detail (smother). They believe humans need rules and controls to become good.

Mencius advocated for a balance between the two: we should neither ignore nor smother human nature. Like growing rice, if we ignore our nature, it withers. If we smother our nature, we will not flourish.

Introducing a 2,400-year-old poem into your next business proposal might pose a challenge (albeit an interesting one). However, Mencius' teachings serve as a compelling treatise on promotion. To grow an audience, we must neither ignore nor smother them. Ignore an audience and they will never grow; smother an audience and they will never return.

Growing Without Pulling

Akin to planting a seed then ignoring it, lack of promotion stifles communication with audiences because messages go unnoticed. The message lies underground motionless, full of potential, yet never breaking the surface. Over time, these dormant ideas rot under the soil, irretrievable, wasted, and unseen.

Lack of promotion affects user experience. We sometimes understate the differences between what is trivial and what is important. We say too much and express too little. Communication requires emphasis. Heaps of content overshadow a single important message, like a field of weeds hiding a single sprout. How can an idea see the light of day? We must fertilize it.

Promotion is fertilizer, although some marketers refute the comparison: one person's promotion is often another person's bullshit. We know it when we see it: "20% off," "Buy one get one free," "Sign up and save." Such promotions have their place, but we sometimes need a lighter promotional hand—with less pulling. After all, we cannot label every button "CLICK ME!" to get a user's attention. Here, we need to design an environment that holds a user's focus by removing the unnecessary to promote the important. Like a gardener cultivating a field, we weed out the distractions and let the vital ideas take root.

---

[2]Norden, Bryan Van. "Mencius." Stanford Encyclopedia of Philosophy. October 16, 2004. Accessed June 08, 2018. https://plato.stanford.edu/entries/mencius/.

Vital ideas must be noticed. A just noticeable difference (JND) is the minimum amount of perceivable difference. Whereas a "5% Off Sale!" goes unnoticed, a "20% Off Sale!" may drive customers to action. The JND is the difference between noticing and not noticing. Promotion affects user experiences by pulling some parts of an experience forward in time—just enough to be noticeable. JNDs work similarly in interface design.

Imagine for a moment that you have never before seen a calculator. You scan the array of buttons—numbers, plus, minus, and the equals sign. Out of curiosity, you start tapping them. You tap the number 2, then the plus sign, and then another 2. You might expect to see "4" upon tapping the equals sign. 2 + 2 = 4. Simple enough, yes? But what would your experience be like if you had never noticed the equals sign? As a first-time user of a calculator, you would likely never realize the utility of the device as a whole. You would never see the result of your efforts. All input, no output. You may think, "This is pointless or broken—perhaps both!"

How often have you thought the same when using an application? You notice a feature but not its benefit. Consider a website's email newsletter sign-up. Sometimes it is a pointless element. "Sign up for our newsletter!" the interface demands. The company may have an excellent newsletter, but such an evaluation can only be made once a person receives it. However, sign-ups require users to supply their email addresses first. Thus, we have a dilemma: how does a user realize a benefit before she takes any steps to achieve it? Answer: spread a bit of fertilizer.

We could show a user an example newsletter. We could highlight its helpful content. We could promise never to spam. These are examples of promotion. We pulled a part of a user's experience forward in time—just enough to make it noticeable. By promoting a benefit, we promote its related features.

Conversely, some companies mistakenly equate prominence with promotion: they splatter big, bold messages across screens, often for reasons unknown even to themselves. They violate their users' experiences with distractions—garish carousels, featured news releases, and useless social feeds. These companies could have instead promoted a single, considered action: buy a product, create an account, or any other goal that establishes a relationship with a user. Because, in the end, that is what promotion must do. We can neither ignore nor smother an audience. We must sow, cultivate, and harvest along the way.

Scarcity

How do you persuade an entire country's population to eat something they do not want? Something so reviled that farmers would not feed it to their livestock. Something so unpopular that it caused riots in Russia. Something so forsaken that clergy called it the "devil's apple."[3] Answer: you create a scarcity.

For centuries, the leaders of European countries sought protections against famine.[4] Wars and widespread crop failures throughout the continent wreaked havoc on agriculture, as well as the societies it fed. At the turn of the 17th century, one full third of the Russian population died of hunger.[5] By midcentury, the majority of Central European countries had fought a continuous series of wars for three decades, killing hundreds of thousands. And, in the century's closing years, famine had spread as far north as Finland. The population of Europe needed nutritious food... something easy to grow... something easy to store... something easy to eat. Behold the common potato.

Originally cultivated in the South American Andes, potato-farming grew across the heartland of Central America. Spanish explorers extended the potato's reach into Europe in the 16th century. Soon after, folklore of the potato's wicked ways took root in the minds of Europeans. Some said potatoes caused leprosy,[6] a few called them the "Earth's testicles,"[7] and nearly everyone refused to eat them.

In 1774, Frederick the Great, King of Prussia, introduced potatoes to his country.[8] As kings often do, Frederick employed the surefire marketing tactic of intimidation. He cajoled. He demanded. He threatened. Yet, this strategy proved unsuccessful. He could neither sell nor give potatoes away. The more Frederick pushed, the more his people resisted. The citizens of Prussia had no desire for spuds. The abundance of potatoes was surpassed only by the soil in which they were planted—to a Prussian, the likelihood of eating either was about the same.

---

[3] Toomre, Joyce Stetson., and Elena Molokhovets. *Classic Russian Cooking: Elena Molokhovets' A Gift to Young Housewives*. Bloomington: Indiana University Press, 1998.

[4] "List of Famines." Wikipedia. June 03, 2018. Accessed June 08, 2018. https://en.wikipedia.org/wiki/List_of_famines.

[5] "Droughts and Famines in Russia and the Soviet Union." Wikipedia. June 05, 2018. Accessed June 08, 2018. https://en.wikipedia.org/wiki/Droughts_and_famines_in_Russia_and_the_Soviet_Union.

[6] Stradley, Linda. "History of Potatoes, What's Cooking America." What's Cooking America. August 11, 2016. Accessed June 08, 2018. https://whatscookingamerica.net/History/PotatoHistory.htm.

[7] Kiple, Kenneth F., and Kriemhild Coneì Ornelas. *The Cambridge World History of Food*. New York: Cambridge University Press, 2000.

[8] "The Legend of the Potato King." *The New York Times*. October 11, 2012. Accessed June 08, 2018. https://niemann.blogs.nytimes.com/2012/10/11/the-legend-of-the-potato-king/.

Citizens needed a staple crop, but all the king's attempts at persuasion rotted in the ground. What was a ruler to do? Frederick needed to create demand for a product that everyone could use, yet nobody wanted. He needed to do far more than grow potatoes: he needed to cultivate desire.

Frederick continued to grow potatoes in his royal fields, but he added a unique, new twist: he added armed guards. He placed his potato fields and harvests under his royal protection. Rather than sell or give potatoes away, Frederick hoarded them. Nearby villagers likely watched royal guards patrol the fields of growing potatoes, glittering like steel-clad scarecrows under the noonday sun. Over time (and perhaps embellished by legend), observers of these highly secured tubers began to want what they could not have. Potatoes were no longer abundant and unwanted: potatoes were scarce and protected by armed guards.

At night, the guards would allow thieves onto the farms. The thieves carried the potatoes back to their own farms for replanting, all with the secret blessing of the king. Planting led to harvests. Harvests filled storehouses. Storehouses fed families. By creating scarcity, Frederick grew much more than potatoes: he cultivated desire.

An experience starts with a perception—good, bad, or indifferent. We often perceive scarcity as an attribute of something valuable, from a diamond adorning a wedding ring to the common potato being placed under armed guard. Scarcity not only quantifies, it also qualifies.

Scarcity by Amount

Tell a person something is scarce, and you have created a scarcity. By defining such an anchor, we create a perceptual contrast between it and any other subsequent data. "One of ten," may convey a scarcity, whereas "one of a thousand" may not. We perceive scarcity through a comparison between what is offered and what is claimed. Eighteenth-century Prussians were offered millions of potatoes; however, no potatoes were claimed. Nothing claimed, nothing scarce. Once the majority of potatoes were claimed, the resulting comparison changed. A careful balance between these two sides creates scarcity while maintaining availability.

Sophisticated e-commerce solutions utilize stock counters to show limited availability. Rather than show 999 of 1,000 available, the availability is hidden until the careful balance is reached between what is offered and what is claimed. The display of "10 left in stock" compels shoppers to buy, because the perception is that far more than 10 were initially offered. As you could imagine, "10 of 10 left in stock" may make a product appear as undesired as the devil's apple.

Scarcity by Time

If you have ever attended an auction, you can appreciate the notion of scarcity by time. Auctions excite audiences by compressing the time between what is offered and what is claimed into seconds. "One hundred, do I have two hundred? Two hundred, do I hear three hundred? Three hundred, going once, going twice..." Offers fly off an auctioneer's lips and onto bidders' checkbooks quicker than you can say, "I've been persuaded by a scarcity marketing technique."

Online auctions persuade and offer much of the same excitement, albeit with less pressure. Their durations last longer than offline auctions and the bidding is frequently automatic. Persuasion comes in the form of time, but wraps itself in the promise of a potentially good deal.

Limited time offers mimic the persuasive effects of auctions. The implicit promise of a limited time offer is a potentially good deal. After all, a limited time offer would not be very compelling if you perceived the deal improving after the time limit expired. Once we doubt such a promise, its persuasive effects become far less effective. Consider the furniture retail stalwart of a going out of business sale.

Scarcity by Exclusivity

The labeling of an item with "special edition" connotes scarcity through exclusivity. It is the highest form of scarcity because, true to its name, its specialness is what delineates between what is offered and what is claimed. It is not merely scarce, but scarce for a reason.

Why more companies do not utilize special editions is perplexing. Who wouldn't wish to buy a special edition over the humdrum, normal edition of a product? You could imagine a special edition thumbtack, jet liner, or tub of butter. All of which are somehow magically imbued with the touch of uniqueness, simply by being called a special edition.

Like limited time offers, the technique may backfire if not handled with care. Special software editions work best as unique branches off the main trunk of an application. Game publishers extend their reach to new audiences through licensed and seasonal editions. A visit to the App Store uncovers not only Angry Birds, but also Angry Birds Star Wars, Angry Birds Rio, and Angry Birds Seasons. Special editions of a main trunk may confuse audiences and are best avoided. The most recent, stable version of software is special enough.

Ask yourself if your creation could benefit from scarcity. Perhaps you release an application to 1,000 initial users, then collect additional email addresses on a waiting list. Premium memberships, per incident support, and limited availability of in-application items are all means to seed curiosity within an audience.

To this day, people still leave potatoes on King Frederick's gravestone—an offering to an effective persuader. Persuasive techniques are often maligned for their manipulative outcomes. However, we should acknowledge the occasionally good outcomes as well. You need not save the world through your creations, but you will often find that scarcity can grow interest in even the most fallow of fields.

Key Takeaways

- When handled appropriately, promotion helps users recognize critical information.

- We must neither ignore nor smother users.

- A just noticeable difference (JND) is the minimum amount of perceivable difference between two pieces of information.

- Scarcity creates a perceptual contrast between what is offered and what is claimed.

Questions to Ask Yourself

- Does an experience effectively communicate its benefits to users?

- Am I smothering users with too much information?

- Can I foster interest in a product or service by reducing its availability?

- Is there an opportunity to create a special edition of a product or service (e.g., a pro version)?

Place

Soviet-era films stirred hearts, indoctrinated the masses, and influenced generations. The films pioneered cinematographic techniques that are still in use today. One technique in particular yielded a surprising effect. Filmmakers discovered that by placing one image before another, an audience's perception of the second image changed. For example, an audience believed an on-screen actor looked hungry after they viewed an image of food. Known as the Kuleshov effect[1] (see Figure 29-1), the technique can be found in everything from Cold War thrillers to modern-day user experiences.

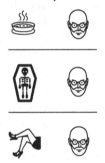

Figure 29-1. Kuleshov effect illustrated in static graphics[2]

---

[1]"The Kuleshov Experiment." Elements of Cinema. Accessed June 08, 2018. http://www.elementsofcinema.com/editing/kuleshov-effect.html.
[2]Mayor, Jems. "Skeleton." Digital image. The Noun Project. Accessed June 7, 2018. https://thenounproject.com/term/skeleton/229927/
HeadsOfBirds "Legs." Digital image. The Noun Project. Accessed June 7, 2018. https://thenounproject.com/term/legs/1154879/
zidney "Soup." Digital image. The Noun Project. Accessed June 7, 2018. https://thenounproject.com/term/soup/1675597/
Yu luck "Persona." Digital image. The Noun Project. Accessed June 7, 2018. https://thenounproject.com/term/persona/623591/

© Edward Stull 2018
E. Stull, *UX Fundamentals for Non-UX Professionals*,
https://doi.org/10.1007/978-1-4842-3811-0_29

Lev Kuleshov exhibited his method in 1918 when he projected several short movie scenes in front of an audience. Kuleshov displayed a bowl of soup then an image of a man's emotionless face. (Audiences thought the man was hungry.) Kuleshov displayed a child in a coffin then an image of a man's emotionless face. (Audiences thought the man was sad.) Kuleshov displayed a woman reclined on a sofa then an image of a man's emotionless face. (Audiences thought the man was full of lust.) In each case, the man's face was the same image, but audiences derived differing perceptions of the man based on what image had preceded him. Hungry. Sad. Lustful. Placement changes everything.

Placement in User Experience

The Kuleshov effect shows us that placement determines context. Previous experiences inform current ones; current experiences inform future ones. Soviet filmmakers understood this well, as do application designers.

Consider an account sign-in. When a person has an account, the sign-in screen may serve as a place to enter a user name and password. When the person does not have an account, the sign-in screen may serve as a place to create one. Same screen, different contexts. This happens any time a screen serves multiple purposes.

A website may serve dozens of contexts: shopping, product research, and career seeking, to name but a few. Like Kuleshov's example, users will derive differing perceptions of this website based on their context. To support these users, we must understand where they came from to anticipate where they wish to go. Perhaps they shop. Maybe they research. Possibly they seek a career.

Individual interface elements are also places, as each is a potential target of a user's attention. Users move from titles to text, from buttons to labels, from lists to links, from target to target. Fitts' law demonstrates why some movements are quicker and easier than others. At the heart of the law lies a simple model: the time and difficulty to reach a target is determined by the size and distance of the target. For example, moving your mouse pointer (i.e., cursor) between two adjacent, average-sized links is quick and easy, whereas moving between two distant, small links is slow and difficult.

TARGET 1 | TARGET 2

vs.

TARGET 1 | TARGET 2

When speed is needed, place targets close together. When distance is desired, make the targets bigger. We see Fitts' law play out every day when typing on keyboards, selecting from menus, and hunting-and-picking buttons and links within an interface.

Place extends to offline experiences, as well. Paco Underhill's book *Why We Buy: The Science of Shopping* describes place through the lens of retail design. Though the book's focus rarely shifts from a brick-and-mortar store setting, the lessons learned are transferable to a multitude of digital user experiences. How shoppers enter and exit a building may affect their ability to understand information: a building's entrance serves as an "outside–inside" decompression zone for visitors, allowing them to adjust to a new context. It is a magnificent model of how to introduce users to a new experience, be it a store, a website, or an app. Since digital experiences can change in microseconds, users often need to pause to take a breath.

In that spirit, let us pause for a moment and discuss the next section: process. Process requires all of what we have covered so far. We will exercise empathy. We will confront authority. We will seek motivation and relevancy and hope our efforts are reciprocated. If there were a fifth "P" in the four Ps (product, price, promotion, and place), it would be process, because process affords you an opportunity to craft the most important user experience of all: your own.

Key Takeaways

- Placement determines context. Past experiences affect current and future experiences.

- Users will derive differing perceptions of an experience based on their context.

- Fitts' law demonstrates that the time and difficulty to reach a target is determined by the size and distance of the target.

- To speed up an experience, place targets close together.

- "Outside–inside" zones allow users to adjust to a new context.

Questions to Ask Yourself

- What was a user doing before an experience?

- Does each part of an experience (e.g., a sign-in screen) address every user's context?

- What will the user do after an experience?

- Do I want to speed up or slow down a particular user behavior?

- How might an experience be handled in a different channel (i.e., online compared to offline)?

Process

Roughly every five years North Korea holds an election. Yes, the same country that kidnaps movie stars, mandates 28 state-approved hairstyles,[1] purges its capital city of all short people,[2] and threatens the world with nuclear Armageddon, also holds elections. Voters approve a single name on a single ballot with a single "yes" or "no" vote. It is a remarkably simple process. Dictatorship can be wildly efficient.

The last election, held in March 2014,[3] resulted in unanimous agreement: with a reported 99% turnout in all 687 districts, the civic-minded inhabitants of the Democratic People's Republic of Korea approved their leader without a single dissenting vote—out of 25 million.[4] But, like many other processes, its potential was never realized by the people who needed it most.

Although we may sometimes act like tiny despots, we often rely on the help of others. A modern-day digital project involves managers, copywriters, designers, developers, subject matter experts, testers, and others. Each person filling these roles serves both as a worker and a collaborator, simultaneously advancing an idiosyncratic agenda and a team's overall mission. Where we succeed is where these two goals converge; when we falter is when they veer off course.

---

[1] @cmsub, Courtney Subramanian. "These Are North Korea's 28 State-Approved Hairstyles." *Time*. February 25, 2013. Accessed June 09, 2018. http://newsfeed.time.com/2013/02/25/these-are-north-koreas-28-state-approved-hairstyles/.

[2] Rundle, Michael. "Kim Jong Il: 18 Strange 'Facts' About The North Korean Leader." HuffPost UK. December 19, 2011. Accessed June 09, 2018. https://www.huffingtonpost.co.uk/2011/12/19/kim-jong-il-18-strange-facts_n_1157276.html?guccounter=1.

[3] Associated Press. "North Korea's Kim Jong-un Elected to Assembly without Single Vote Against." *The Guardian*. March 10, 2014. Accessed June 09, 2018. https://www.theguardian.com/world/2014/mar/10/north-koreas-kim-jong-un-elected-assembly-vote-against.

[4] "The World Factbook: KOREA, NORTH." Central Intelligence Agency. June 04, 2018. Accessed June 09, 2018. https://www.cia.gov/library/publications/the-world-factbook/geos/kn.html.

You may be a high-powered executive, a college intern, or someone in between; yet, our day-to-day experiences create patterns of surprisingly similar struggles and triumphs. We all face disappointment and seek inspiration. We see confusion turn into clarity and witness progress dissolve into chaos.

You might think we would be better at recognizing these patterns. After all, the better part of our waking lives is dedicated to work—90,000 hours, on average.[5] Perhaps we do not recognize these patterns because are too busy to notice; yet, according to a recent Bureau of Labor Statistics report,[6] Americans work more or less the same number of hours per week as we did in 1976. It could be because we work in increasingly varied environments; although, many of us work in nearly homogenous socioeconomic bubbles.[7] Or, maybe we do not recognize these patterns because we have simply stopped trying. If we wish to revolutionize our work, we need a process.

This section of the book helps you create a process that suits your particular needs. It covers the creation, management, and execution of digital projects. We will discuss what usually works and what often does not, detailing a full range of subjects from Agile to user testing.

In the end, we judge a process by the value it creates: project clarity, team happiness, and financial gain. Perfection is not our goal. We wish only to improve the experience of our products, our users, and ourselves.

Let the revolution begin.

---

[5]"Table B-2. Average Weekly Hours and Overtime of All Employees on Private Nonfarm Payrolls by Industry Sector, Seasonally Adjusted." U.S. Bureau of Labor Statistics. Accessed May 28, 2018. http://www.bls.gov/news.release/empsit.t18.htm.
[6]"Women in the Labor Force: A Databook (2010 Edition)." U.S. Bureau of Labor Statistics. March 16, 2011. Accessed June 09, 2018. https://www.bls.gov/cps/wlftable21-2010.htm.
[7]Murray, Charles. "Did You Grow up in a Bubble? These ZIP Codes Suggest You Did." PBS. April 07, 2016. Accessed June 09, 2018. https://www.pbs.org/newshour/economy/did-you-grow-up-in-a-bubble-these-zip-codes-suggest-you-did.

Waterfall, Agile, and Lean

White lights give way to virtual sunrises deep within Norway's longest tunnel. Located between the southwestern municipalities of Lærdal and Aurland, 15 miles of underground roadway gently unfurls beneath the unspoiled landscape of fjords and mountainsides.[1] The Lærdal tunnel stands as an engineering triumph. Two and a half million meters of Precambrian gneiss rock were drilled, exploded, excavated, and removed. Two hundred thousand bolts[2] secure its walls. However, the tunnel's most remarkable attribute is its surprising psychology—a psychology befitting any project.

Equally distanced throughout the tunnel, three large caverns strain under a thousand meters of chiseled rock, the foundations of ancient churches, and the plunge pools of cold waterfalls. Each cave stretches wide enough to contain a small theater and its walls are illuminated in bright tones of blue and orange (see Figure 30-1). This cascade of color emulates a morning's dawn, awakening weary drivers and comforting claustrophobic passengers during their 20-minute journey.

---

[1] Anhie. "Laerdal Tunnel." Atlas Obscura. June 04, 2013. Accessed June 22, 2018. https://www.atlasobscura.com/places/laerdal-tunnel.
[2] THE ENGINEERING.com. "Laerdal Tunnel ENGINEERING.com." Engineering.com. October 13, 2006. Accessed June 22, 2018. https://www.engineering.com/Library/ArticlesPage/tabid/85/ArticleID/60/Laerdal-Tunnel.aspx.

© Edward Stull 2018
E. Stull, *UX Fundamentals for Non-UX Professionals*,
https://doi.org/10.1007/978-1-4842-3811-0_30

Figure 30-1. Lærdal tunnel's illuminated caves[3]

Scandinavian research organization, SINTEF, designed the Lærdal tunnel's lighting based on proposals by psychologists, artists, and designers.[4] Although a driver's journey does not require the trio of light shows, it is certainly improved by their addition. The tunnel remains the same length, with or without the momentary reprieves. This design shows that we can improve any experience—even those engineered to be efficient as possible. Moreover, it demonstrates what is first needed in order to design any project: a plan.

Your Project: The Mountain

Imagine your project as a mountain. Your team stands at its base and wishes to reach the other side. They swing their picks and shovels and dig into the work.

Although we have many methods to organize these efforts, three reign supreme: Agile, Lean, and Waterfall. Each concept can cover anything from manufacturing products to managing startups; however, we will talk about

[3]Bye, Rob. "Tunnel, Road, Underground and Dark HD Photo by Rob Bye (@robertbye) on Unsplash." Unsplash. September 26, 2017. Accessed June 09, 2018. https://unsplash.com/photos/ZKbwjlbwleI.
[4]Fisher, Donald L. *Handbook of Driving Simulation for Engineering, Medicine, and Psychology.* Boca Raton: CRC Press, 2011.

them as they pertain to digital projects. They can govern a part of a digital project or the entire thing. We will discuss all three. Let us tackle Waterfall first.

You are likely accustomed to Waterfall projects (see Figure 30-2). Here, we organize a project into discrete steps, such as research, design, development, and testing. The next step does not begin until the previous one has ended. You start by planning your course through the mountain. Next, you draft blueprints and design the supports. You ready the picks and shovels. Once preparations are complete, the digging begins.

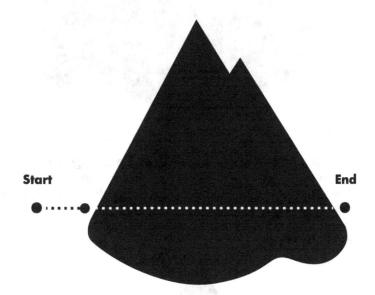

Figure 30-2. Waterfall projects tend to be a non-overlapping, discrete divisions of labor

Using Agile is like carving a tunnel through a mountain 10 feet at a time (see Figure 30-3). At the end of each 10-foot segment, you meet with your team and evaluate where you all stand. You stare at the chiseled walls of the tunnel and evaluate the project's progress. You see that the team chipped away at some features and perhaps watched cracks form within others. Maybe a few sprung leaks. Everyone decides how to move forward, and another 10-foot dig begins. If you veer off course, you can correct it in the next segment. Through repeated segments, or "sprints," the team reaches daylight at the other end.

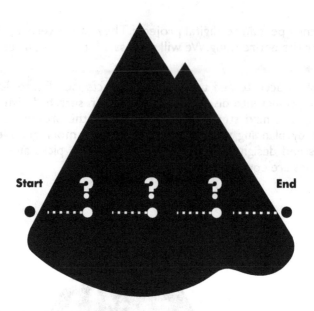

Figure 30-3. All work within an Agile project is divided among sprints

With a Lean project, you choose the shortest route through the mountain—the "minimum viable" route (see Figure 30-4). You survey the landscape to avoid the hard features and potential cave-ins. You include only what is vital. Although the best path may be a longer one, the team forgoes an ideal solution in favor of speed. The team carves away at the project and quickly reaches an outcome on the other side.

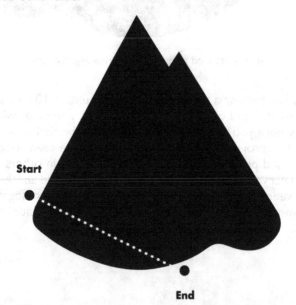

Figure 30-4. Lean projects focus on the creation of a minimum viable product (MVP)

On the surface, Waterfall, Agile, and Lean sound equally well suited to tackling a project. But, like tunneling through a mountain, all projects encounter obstacles along the way. We must plan for detours. It is easy to convince ourselves that the wrong course is the right one when we are stuck under a mountain of work. Let's dig further into each one, starting with Agile.

Agile

Agile traces back to software development methodologies created in the 1960s. In the decades that followed, the strategies continued to evolve. By 2001, *The Agile Manifesto*[5] had introduced a cohesive concept of Agile based on lofty and inspiring goals: working software over documentation, flexible schedules over a rigid timeline, and collaboration over antagonism.[6] Sounds pretty great, doesn't it?

The Agile process and its varieties, such as Scrum and Kanban, have become the norm in large-scale development efforts. The quick path to functioning software does much to curb the apprehensions of today's executives.

Development tasks lend themselves to Agile. Coding is a complex and artful task; however, code either works or it does not. If it meets performance goals then, by most accounts, the code is viable. Using Agile, you can build, test, and deploy almost anything, as long as it fits within a sprint. A sprint is any time-boxed interval, though most are a few weeks in length. You determine the work. You complete it. You test it. You deploy it. You move on to the next sprint. After several sprints, you arrive at a fully functioning product. This model works so well for some organizations that Agile has extended into Agile product development, Agile marketing and—what we will focus on next—Agile UX.

In several environments, Agile UX may be a perfect solution. We split a set of tasks into sprints and quickly see results. Our UX and development deliverables align. Everyone uses a common vernacular. The team digs 10 feet into the mountain, looks around, high-fives, and plans the next 10-foot increment. Tunnels get built.

---

[5]"Agile Manifesto for Software Development." Agile Alliance. November 15, 2017. Accessed June 22, 2018. https://www.agilealliance.org/agile101/the-agile-manifesto/.
[6]"Agile Software Development." Wikipedia. June 08, 2018. Accessed June 09, 2018. https://en.wikipedia.org/wiki/Agile_software_development.

But this immediacy is also where cracks begin to form within an Agile project. Although speed may be a virtue, it misses the small issues that, if left unaddressed, may grow into tremors, crumble the support for your project, and bury you alive. We make tradeoffs. Rather than wait for a research study, we interview only a handful of stakeholders. Rather than wait to build wireframes, we advance to a prototype. Rather than wait for conclusive testing results, we repair as we receive feedback. We trade clarity for speed, and contemplation for immediacy. These tradeoffs can be compelling, but they are also why Agile UX projects sometimes fail.

Why would an Agile UX project fail? After all, many UX research and design processes can be grouped into tasks, and these tasks are frequently iterative. Furthermore, UX deliverables are notoriously document-heavy, and Agile UX promises to lighten this load. People would rather play with a functioning prototype than trudge through detailed documentation. And lastly, people want software now, not months from now. It is hard to argue against any of those points. But I will.

Problems arise from the UX tasks themselves: several are linear and build upon one another. Experienced team members may appreciate the first stages of user experience design, but these early activities are intangible to novices. Research may appear too slow. Personas may appear too silly. Flowcharts may appear too abstract. Wireframes may appear too fastidious. The early steps of a UX process may appear to be sedentary: people talk about building a tunnel, but nobody is digging.

We must realize that digging a tunnel into a mountain and digging a tunnel into a volcano may look very much the same, at first. We do not notice the difference until we have reached the middle (see Figure 30-5). Some projects combust without effective planning and research. Planning and research reveal their worth over time. Conversely, we recognize their absence when it is already too late, after having squandered vital weeks on a flawed prototype. We find ourselves running, screaming, and searching for a project's exit. With scorched eyebrows and charred egos, we promise to find a better way next time.

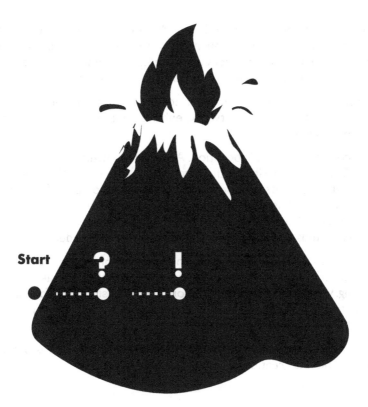

Figure 30-5. Agile projects may lead to unpleasant surprises[7]

Much of UX design and research is about planning what to do—what will be experienced by users. UX design and research is the precursor to visual design and development. It creates the blueprint that charts a course through the mountain. It avoids the lava.

A Tale of Two Ideas

You will likely face one of two possible scenarios when working on an Agile UX project: in the first, you want to add or edit the features of an existing product; in the second, you wish to create an entirely new experience.

Maintenance projects tend to fall into the first camp. After an application is first built, ongoing upkeep becomes the norm. New features are added. Some are edited. Others are removed. These incremental additions, modifications,

[7]Iga "Flame." Digital image. The Noun Project. Added icon to mountain illustration by Edward Stull. Accessed June 7, 2018. https://thenounproject.com/term/flame/1745276/

and deletions often work well within an Agile environment. You collaborate with your team and discuss the nuances of features: the smallish tweaks and tiny enhancements to an experience. We determine which tasks to fit within the next sprint. The team agrees on where to dig, and they start digging.

When creating an entirely new experience, we perform many of the same behaviors: collaborate, discuss, tweak, enhance. But, we must also add one more behavior: approve.

Approval conflicts with collaboration. Agile prides itself on collaboration. It gains its efficiency by sacrificing the formality of communication: the approval of ideas. Agile supplants this formality with the promise of changeability: "whatever we create is only an iteration." Approvals be damned! Yet, when creating something anew, we must often obtain approvals to safely move forward. Teams must be convinced. Clients must be persuaded.

Agile projects sometimes crack under the pressure of unapproved ideas. They erode a project's support. Combined with the tradeoffs we make in research and planning, we either veer off-course or come to a standstill. To find our way out of the project, we are forced to chip away at small ideas. Small ideas fit sprints. Big ideas move mountains.

Like many processes, Agile UX has its place. Often, that place is a product's maintenance, not its creation.

Lean

Lean UX is an evolution. It finds a halfway point between Waterfall and Agile. An excellent book on Lean UX is Jeff Gothelf and Josh Seiden's *Lean UX: Applying Lean Principles to Improve User Experience*. The book contains a wealth of helpful tips on everything from team dynamics to prototyping.

At the core of Lean UX is the "minimum viable product"—an MVP. To explain this concept, let's return to our example of digging a tunnel through a mountain.

A tunnel is a big project. It requires considerable time and resources. Your team needs an assortment of picks, shovels, and perhaps a few sticks of dynamite. All these resources cost money. We chisel, excavate, and detonate. All these activities take time.

With every project, we risk wasting time and resources. We chart the wrong course. We run into impenetrable obstacles. We dig into the wrong mountain.

The brilliance of Lean UX is its lack of ambition. If Lean UX had a rallying cry it would be "Let's... not!" In essence, the minimum viable product is the shortest path to success. We reduce risk by avoiding large expenses of time and resources. It is akin to digging the shortest route through the mountain.

The route we take is not necessarily ideal, but by completing even this short path, we gain new knowledge. Along the way, we learn about potential pitfalls and uncover veins of gold. But our greatest learning comes from reaching the other side of the mountain. For the first time, we see what awaits us. We may learn that the resulting landscape is not worth the effort—best to stop now. We may learn that an ideal destination is nearly in reach—best to keep on digging.

You will find Lean UX practices within small startups and large corporations. An MVP serves as a proof of concept. More than a prototype, it demonstrates the crucial features of a product—not all, just the ones that make the product viable. For example, a map app should display maps. An auction site should accept bids. A banking kiosk should provide account balances. Once we achieve the minimal viable product, everything else becomes elective.

Determining what is crucial and what is elective challenges even the most experienced of teams. What should be included? Whereas an application's stability may be viewed as crucial, an application's aesthetics may not. Does an app require an optimum user experience? I think so, but you may feel differently. A Lean UX project includes only the necessary. It is practical. It is realistic. And, it is often rather boring.

An MVP rarely stirs the heart—it is the minimum, after all. Much of what compels users are the product's details: the micro-interactions, the small gestures, and the tailored experiences. Although the minimum gets the product out your door, it does not necessarily get it into a customer's.

An MVP provides us with a start: a glimpse of the other side of the mountain. We can either abandon our effort or keep on digging. Our goal is to reach an ideal state—a product that not only offers the minimum but also all the electives that make an experience optimal and enjoyable.

The key to reaching this promised land is buried somewhere deep within your project. You only need to look. Start with an MVP. Add, edit, and delete until the experience is so ideal, so perfect, that no other path through the mountain would be as gratifying.

A Bit of This, a Bit of That

As we conclude this chapter about Waterfall, Agile, and Lean, we should reflect on our earlier discussions about being human. Because, over time, we realize that any discussion of process is actually about people.

We need not divide ourselves into separate tribes of project managers, designers, or developers. Our commonalities are the key to working with one another. All of us are remarkably similar: we wish to succeed, and we fear restrictions. Yet, if we are honest with ourselves, we must admit that we find comfort in imposing limitations on others. To do so gives us a sense of predictability and safety.

Even those who say, "we do not need a process," are often the first people to demand guidance at the first sign of difficulty. Thus, we must acknowledge this need for predictability, for people can erect walls as easily as they can dig tunnels. A good process provides a means of protection, as well as action.

Is there a best process, one that works every time—a secret to combining inspiration, expertise, rigor, profit, and personal satisfaction? The honest answer is no. We live in a world governed by happenstance: economies thrive or take downturns, clients succeed or go bankrupt, users become loyal or defect to our competitors. Too many processes attempt to satisfy all people, tasks, and timelines. A process tailored to social marketing may fail when developing enterprise software. One befitting seasoned managers may fail when applied to recent college graduates. Your process has little-to-no effect outside your office's walls.

Where does this leave us? Though no process is perfect, a common thread runs through many successful ones. You can leverage the speed of some and the clarity of others, building a process suited to your particular circumstances.

We need not choose only from among Waterfall, Agile, and Lean. These methodologies can improve work, but there is more than one way to tunnel through a mountain. You could research a project using Waterfall, design a product using Lean, and end your development with Agile… or vice versa… or by applying any other combination.

It would be naive to think that all situations can be covered by a single process. Choose what works best for you. After all, you are the one doing the digging.

Key Takeaways

- Agile prioritizes working software over documentation, flexible schedules over a rigid timeline, and collaboration over antagonism.

- Early UX activities, such as research, are often intangible to novices.

- Research reveals its worth over time.

- Approval conflicts with collaboration.

- Agile projects sometimes fail because of unapproved ideas.

- Agile methodology favors a product's maintenance, not its creation.

- Lean UX focuses on the "minimum viable product"—a testable proof of concept.

- Any discussion of process is actually about people.

- A good process provides a means of protection as well as action.

- Build a process suited to your particular circumstances.

Questions to Ask Yourself

- What are each of my team member's needs?

- What approvals are necessary to complete the project?

- What are the known obstacles in my project?

- Have I included my team in user research?

- How can I create a parallel research track unconfined by a sprint schedule?

- Am I working on the best ideas or ideas that simply fit into a sprint schedule?

- What UX documentation is required for each team member to do her or his job?

- Do remote team members require additional UX documentation to compensate for communication challenges (e.g., disparate time zones, nonnative languages, or varied national holidays)?

- Will any team member be unavailable during the project (e.g., planned absences, conflicting project schedules, or new-employee onboarding)?

- Can my project use a combination of Agile, Lean, and Waterfall methodologies?

Problem Statements

The Hutzler 571 slices bananas. Shaped like a banana, the multi-blade tool has garnered over 5,000 Amazon reviews.[1] Most are satirical. One reviewer writes, "It saved my marriage," another, "It works better than a hammer." However, like many products, the problem it solves is not always immediately apparent.

You may wonder what problem a banana slicer attempts to solve. After all, you can easily slice a banana with a knife, a fork, or a spoon.

Surprisingly, several of its 5,659 reviews appear to be serious. The Hutzler 571 (Figure 31-1) supposedly works wonders when preparing bananas for food dehydrators. It creates uniform slices, which means uniform drying times. Dehydrators may take hours, undercooking thick pieces and overcooking thin ones. The Hutzler 571 solves that problem. Kids enjoy using the tool, too. So, if you have children, frequently dehydrate bananas, and have a whole bunch of time to waste, you might be thrilled by the Hutzler 571.

---

[1]Hutzler 571 Banana Slicer. Amazon. Accessed June 7, 2018. https://www.amazon.com/Hutzler-571-Banana-Slicer/dp/B0047E0EII/.

© Edward Stull 2018
E. Stull, *UX Fundamentals for Non-UX Professionals,*
https://doi.org/10.1007/978-1-4842-3811-0_31

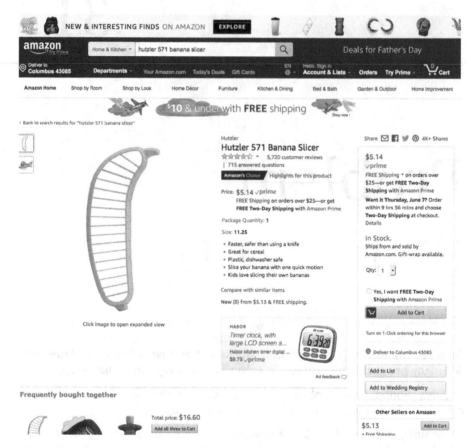

Figure 31-1. Amazon's product page of the surprisingly useful Hutzler 571 Banana Slicer[2]

If Hutzler had described the problem, the company might have written:

> The Hutzler 571 is a kitchen tool. Our growing market of banana-slicing enthusiasts demand a better way to prepare bananas for dehydration. The tool helps customers cut bananas into uniform slices.

Yes, the example is silly, but it represents how stating a problem can illuminate its strengths and weaknesses. Stating the problem helps us design everything from five-dollar kitchen tools to million-dollar digital products.

Before a user clicks, before a website launches, before a proposal is written, we talk about a project. Such discussions affect user experience, because design starts when debate begins.

---

[2]Hutzler 571 Banana Slicer. Digital image. Amazon. Accessed June 7, 2018. https://www.amazon.com/Hutzler-571-Banana-Slicer/dp/B0047EOEII/.

We debate broad categories of understanding: what will we create, why should we attempt it, how can we achieve it? A problem statement frames the answers, reducing all possibilities to a select few: what, why, and how.

Defining What

The philosopher Bertrand Russell said, "The greatest challenge to any thinker is stating the problem in a way that will allow a solution."

How do we state a problem for a digital project? It may originate from a Waterfall, Lean, Agile, or Design Thinking process. It could describe almost anything, covering a wide range of apps, websites, and kiosks. Moreover, a digital project is more than its form; for example, a website could be anything from a personal blog to Amazon.com. We must give it a specific frame.

Imagine you work with Acme Fruit Company. You could describe a project for the company with the following statement:

> Acme Fruit Company will create an e-commerce website to sell fruit baskets.

Though this description is limited, it gives us a specific frame through which to view the project. It tells us something about what we intend to create— and what we do not. The project will be an e-commerce website. We might expect such a website to include a catalog of products, a shopping cart, and so on. It likely does not manage your email, maintain your photo albums, or offer you a publishing platform.

We further frame our project by describing its purpose.

Defining Why

Although a project may exist for several reasons, we want to highlight its primary objective. Doing so focuses our efforts, ensuring our team understands the reasoning behind our problem statement.

> Our website must differentiate Acme Fruit Company within a crowded marketplace.

This statement indicates our primary objective is to differentiate the company.

Defining How

Accompanying our what and why, we describe how. The "how" addresses our primary objective.

> By providing an optimum user experience, we will surpass our competitors' similar offerings.

Putting It All Together

Our problem statement describes what, why, and how we intend to solve a problem.

> Acme Fruit Company will create an e-commerce website to sell fruit baskets. Our website must differentiate Acme Fruit Company within a crowded marketplace. By providing an optimum user experience, we will surpass our competitors' similar offerings.

Now a team can debate and discuss the merits and pitfalls of such an argument:

- Is this a problem worth solving?
- What is the best medium (e.g. website, app, kiosk) to achieve our goals?
- Is the primary purpose to sell products or differentiate?
- Is price the determining factor in a user's purchase decision?
- What do users dislike about buying fruit online?
- Are there untapped markets the competition does not serve?
- How do we define an optimum user experience?
- ...And countless others

Each answer reshapes a project, pushing and pulling the boundaries of our understanding. The problem statement may be made irrelevant by a project's end. Yet, it serves to elicit debate, uncovering gaps in our knowledge. Are there fundamental misunderstandings about the project? Do business objectives conflict? What research might be needed? If we catch our assumptions early, we avoid costly stumbles and last-minute slip-ups.

Key Takeaways

- A problem statement serves to elicit debate by defining the what, why, and how about a project.

- Problem statements catch assumptions, misunderstandings, and gaps in knowledge.

- A problem statement may be made irrelevant by a project's end.

Questions to Ask Yourself

- What is the problem I wish to solve?

- Have I stated the problem clearly enough for another person to debate its assumptions and conclusions?

The Three Searches

Why would anyone buy a pair of $300 sunglasses?

Several years ago, my agency was approached by a high-end sunglasses manufacturer. They wanted to redesign their website. Coincidentally, I had a strong bias against expensive sunglasses for years. They had always seemed wildly unnecessary. After all, chances were I would just lose them. My sunglass cost just $19, and a comparable replacement could always be found rotating on a drugstore spindle.

So how do you confront your own, known biases at the beginning of a project? Easy. Google it.

We sometimes forget what a fast and powerful research tool Google search can be. It is fast, free, and informative. It is history's transcript, retrievable in a few keystrokes.

When researching via Google, the three most important words are "news," "technology," and "vs." Append these terms to any subject you want to learn about and you will receive a wealth of valuable information in return. For example:

> "sunglasses news"
>
> "sunglasses technology"
>
> "sunglasses vs."

Almost every industry generates news, discussions about its related technology, and some form of controversy and competition.

© Edward Stull 2018

E. Stull, *UX Fundamentals for Non-UX Professionals*,
https://doi.org/10.1007/978-1-4842-3811-0_32

Searches for "sunglasses news" provides a detailed analysis of companies, products, events, and trade shows (see Figure 32-1). You discover forums in which customers praise and complain about a company and its competitors. It gives you a knowledge base from which to draw when talking about, writing about, and presenting on the subject.

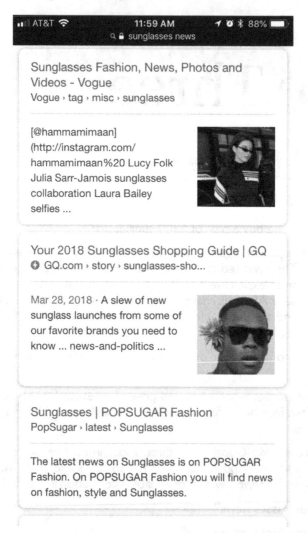

Figure 32-1. Search for research term + "news"[1]

[1]Google search results for "sunglasses news". Digital image. Google Search. Accessed June 07, 2018. https://www.google.com/search?q=sunglasses+news.

Searches for "sunglasses technology" return results about how sunglasses companies differentiate their products through complicated optics, innovative material design, and advanced manufacturing techniques that all factor into developing superior—albeit pricey—sunglasses (see Figure 32-2).

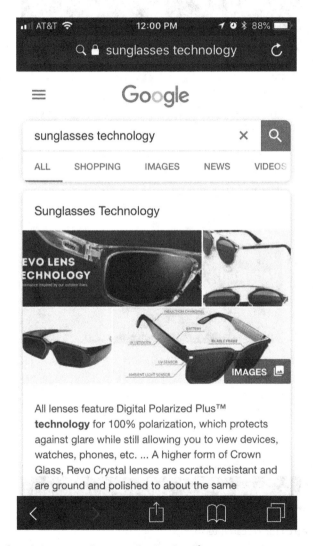

Figure 32-2. Search for research term + "technology"[2]

---

[2]Google search results for "sunglasses technology". Digital image. Google Search. Accessed June 07, 2018. https://www.google.com/search?q=sunglasses+technology.

Searches for "sunglasses vs." lists the ongoing debates between pricey or cheap, darker or lighter, yellow or blue, clip-ons or transitions, Ray-Ban or Oakley, polarized or non-polarized (see Figure 32-3).

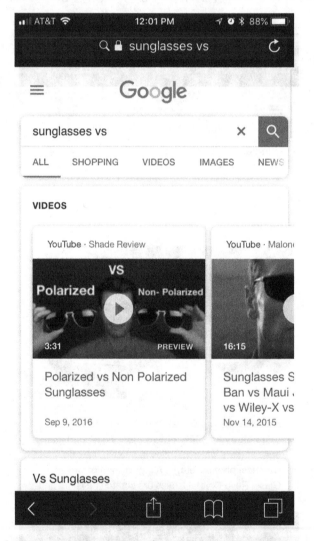

Figure 32-3. Search for research term + "vs"[3]

---

[3]Google search results for "sunglasses vs". Digital image. Google Search. Accessed June 07, 2018. https://www.google.com/search?q=sunglasses+vs.

You learn how high-quality lenses affect polarization. They reduce eyestrain, make colors more vibrant, and eliminate glare. Boaters and anglers love them. Polarization takes the glare off of sunlit water, which is caused when the number of lumens exceeds your eyes' ability to absorb the light (usually around 4000+ lumens).

You read about how high-end sunglasses are often made with memory metals and lightweight alloys, which helps avoid cracking and breaking. Cheap pairs of sunglasses are press-molded with inexpensive polycarbonates and rigid metals.

An hour of Googling shines a light on unfamiliar subjects, revealing insights that may go unnoticed in the early days of a project. It offsets your bias with objective information. And although such research pales in comparison to formalized studies, it is far better than being left in the dark.

Key Takeaways

- Searches for a subject plus "news" return results about related companies, products, events, and trade shows.

- Searches for a subject plus "technology" return results pertaining to the underlying scientific, industrial, and commercial applications of a subject.

- Searches for a subject plus "vs" return results about debates within a subject area.

Questions to Ask Yourself

- What can I learn about my research target through a simple Google search?

Quantitative Research

Life aboard an 18th-century British sailing ship left much to be desired. Salt caked your clothes. Rats shared your food. Hard, sun-baked days retreated into cold, damp nights. You could fall off a foremast, blow over a bow, drown in the deep, succumb to scurvy, or be vanquished by venereal disease. All the while, you counted. You counted the dawns at sea and the stars at dusk. You counted knots, fathoms, and degrees. Ocean crossings became makeshift research studies, where sailors quantified the lengthy distances between ports and the even longer durations between paychecks.

Naval voyages could take months.[1] To discourage sailors from deserting, ship's captains often suspended sailors' wages. The promise of future earnings encouraged love-struck seamen to re-board their ships after long nights of frolicking in foreign harbor towns.

The 1790 diary of George Hodge[2] chronicled his career in the Royal Navy, during which time he waited 17 years to be paid his full wages. In the interim, he and his fellow sailors were given a "tot," a daily ration of rum. Both buyers

---

[1]Henderson, Tony. "Rare Insight into Life under Nelson." *Journallive*. June 25, 2013. Accessed May 28, 2018. http://www.thejournal.co.uk/news/north-east-news/rare-insight-life-under-nelson-4500107.
[2]Henderson, Tony. "Rare Insight into Life under Nelson." Journallive. June 25, 2013. Accessed May 28, 2018. http://www.thejournal.co.uk/news/north-east-news/rare-insight-life-under-nelson-4500107.

© Edward Stull 2018
E. Stull, *UX Fundamentals for Non-UX Professionals*,
https://doi.org/10.1007/978-1-4842-3811-0_33

and sellers valued rum. Its value lay in its alcohol. Unscrupulous sellers would dilute a cask of rum with water, reducing both its alcohol content and its value. In response, buyers devised ingenious ways to test rum's quality prior to purchasing it.

The phrase "keep your powder dry" originates from a warning issued to sailors and soldiers.[3] Once gunpowder becomes wet, it will not ignite. Muskets will not shoot. Cannons will not fire. This is true in all cases save one: gunpowder will ignite in a mixture of water and alcohol, but only when the percentage of alcohol is high enough to counter water's extinguishing effects.

Sailors would sprinkle gunpowder over a small pool of rum and then attempt to set it ablaze. Watered-down rum soaked the gunpowder and would not ignite. It fizzled, instead. But at 57% alcohol,[4] magic happened: the rum burned. A brilliant flash indicated a high percentage of alcohol. Rum that burned gave sailors proof of its quality—the rum was "proofed." This age-old measurement is why we see liquor bottles labeled with their proofs, even today.

Our present-day means of measuring things are more accurate, but the goal is the same: we measure to prove. Like untrusting sailors testing a cask of rum, we wait to see the flash or fizzle before we declare our success or failure.

We now swim in a sea of data. Quantitative research provides us with a means to navigate it. It measures the world through numerical and statistical analyses. It reports budgets, records populations, and measures speeds. What is the average cost of a U.S. aircraft carrier? Where are women-owned firms flourishing? How long does it take for users to check out? On the surface, such data denotes little information other than numbers. But further analysis uncovers additional insights. Soaring budgets may signal a rising commodity market. Successful economic zones may indicate an advantageous tax policy. Long checkout times may reveal problems with a website's shopping cart.

Each measurement quantifies data and shapes our research, proving our success or failure. Where we once guessed people's behavior, we now can track their every click, tap, and swipe. Yet, research looks backward; we see the wake of the ship, but never what lies ahead. We cannot predict the future with certainty, but we can measure which direction the wind is blowing.

---

[3] Hayes, Edward, and William Kenealy. *The Ballads of Ireland.* Boston: P. Donahoe, 1856.
[4] "Alcoholic Proof." Princeton University. Accessed May 28, 2018. http://www.princeton.edu/~achaney/tmve/wiki100k/docs/Alcoholic_proof.html.

Significance

When we delve deeper into quantitative research, we discover that what we are really talking about is significance. Which data aides our decision making and which are merely paper and pixels? A researcher can endlessly record and analyze the world—but to what end? For quantitative research to be useful, it must be practically and statistically significant.

Before you jump overboard, know that we will only skim the surface of statistics here. We will cover the basics while avoiding the details that make math professors rejoice and grad students cry.

Let us start by defining a few terms.

A *population* is the entirety of a data set, be it a population of English sailors, flying fish, or rum barrels. A population includes every sailor, fish, or barrel—not just the big ones; not just the small ones; not just the ones we want to include. Every single one.

A *sample* is a subset of data collected from a population (see Figure 33-1). For example, you could collect a sample of rum by pouring a cup from several barrels.

Figure 33-1. Comparison of data set[5] and sample[6]

---

[5]Alex Muravev "Barrel." Digital image. The Noun Project. Accessed June 7, 2018. https://thenounproject.com/term/barrel/957656/
[6]Natasha Fedorova "Beer." Digital image. The Noun Project. Accessed June 7, 2018. https://thenounproject.com/term/beer/3454/

A *statistic* is a number that summarizes data in a sample. Say we had 10 cups of rum in our sample, ranging from 70 to 80 proof. Our sample might include proofs 70, 71, 71, 71, 75, 76, 77, 77, 78, and 80. To calculate the average, we add all values and divide by the number of values (see Figure 33-2).

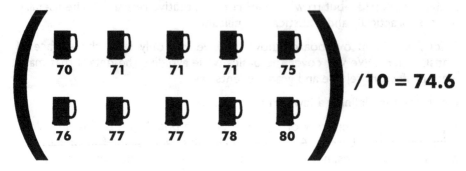

Figure 33-2. All values divided by number of values: (70 + 71 + 71 + 71 + 75 + 76 + 77 + 77 + 78 + 80) / 10 = 74.6[7]

A sample's average alcohol proof is a statistic (e.g., 74.6 proof).

Good statistics are generalizable, meaning the statistic can be used to infer conclusions about an entire population. We say the average alcohol proof of a few cups of rum represents the average alcohol proof of all rum barrels. Generalized statistics are not infallible; they do not always lead to exact matches when extending our research across an entire population. Our sample may indicate an average alcohol proof of 74.6, but a few rum barrels might be watered down, while others might put hair on your chest.

Reliability describes how often a test produces similar measurements under similar conditions. Testing the height of barrels is reliable, as barrels tend to stay the same height over time. By contrast, rum's color is not reliable, because its color fluctuates depending on any number of factors, such as the rum's age and its means of storage.

Validity signifies the accuracy of research. From overall conclusions to individual measurements, we want all research efforts to be valid. Truth be told, lighting rum on fire is not an accurate means to measure alcohol content. A modern-day hydrometer would provide a much more valid measurement. But what fun is that?

[7]Natasha Fedorova "Beer." Digital image. The Noun Project. Accessed June 7, 2018. https://thenounproject.com/term/beer/3454/

Luckily, we have many software tools to analyze populations, samples, and statistics. They do much of the work for us. However, knowing how to analyze data allows us to interpret the resulting information and recognize if it is reliable and valid.

We will discuss issues that affect reliability and validity in the next chapter. For now, let us touch on the primary cause of quantitative research problems: sampling bias. Here we make the error of selecting a non-random sample, thereby affecting our ability to generalize subsequent research findings. Perhaps we select only the rum barrels stored below deck and do not account for the barrels slowly evaporating under the hot sun. Our sample would not be representative of the population. The same may occur when conducting surveys and other quantitative research. We inadvertently select only those people who wish to respond, skipping those who are too busy, uninterested, or unable to answer our inquiries. We miss busy moms, apathetic teens, older adults, non-English speakers, low-income audiences, and people with disabilities.

Assuming we can avoid the perils of sampling bias, we have several ways to collect quantitative data. We collect it through polls, questionnaires, surveys, A/B tests, web analytics, and search logs. Though the methods differ, each attempts to describe data.

Consider the following example:

- 1,000,000 users visit the Fishes'R'Us website every year (population)

- 100,000 users visited the Fishes'R'Us website last month (sample)

- 3,000 users from the sample completed their purchase (statistic)

If we were to measure this sample of 100,000 taken from a population of 1,000,000, you could say that the website converted 3% of its visitors last month. The conversion rate is a statistical mean, averaging all visits from within the sample. Some visits led to a purchase. Many others did not.

A 3% conversion rate is quite good, but what if we wanted to improve it? We could experiment with a free shipping incentive. Our hypothesis: By adding free shipping, we will increase our conversion rate. So, we could run an A/B test, which is a simple comparison of two variants.

Version A: (No free shipping)

- 50,000 users are NOT offered free shipping (sample)

- 1,500 users complete their purchase (statistic)

Version B: (Free shipping)

- 50,000 users are offered free shipping (sample)
- 2,500 users complete their purchase (statistic)

In Version A, our sample returned results that align with those in our previous test—still 3%. No surprise. With Version B, the website had a whopping 5% conversion rate, indicating a correlation between free shipping and conversions. If our test was reliable and valid, we could infer the incentive would be equally as compelling across the entire population.

Yet, quantitative research only discovers what happened; it does not explain *why* something happened. It describes and implies. Its findings are neither complete nor certain. We are not fishing from a barrel.

Correlation and Causality

In the late, cool evening of April 14, 1912, lookouts on the deck of the steamship RMS Titanic spotted an iceberg off their starboard bow. Alarms were rung, orders were issued, and engines were reversed. We know the rest of the story. However, you may be surprised to learn about a remarkable coincidence, and what it shares with software design.

Among the Titanic's survivors was a stewardess named Violet Jessop.[8] In her 24 years, she had already endured much before the ship's sinking, including what must have seemed like a warm-up act to her Titanic voyage. Through an unfortunate stroke of luck, she had just seven months earlier been a crewmember on the doomed RMS Olympic, which nearly sunk off the Isle of Wight as the result of a collision with the warship HMS Hawke.[9] Violet Jessop must have felt a palpable sense of déjà-vu to once again find herself on board an ill-fated vessel when the hospital ship Britannic struck a mine and sank into the Aegean Sea. To survive one maritime disaster is harrowing. Two is unusual. But three is remarkable.

You might think that Violet would reconsider her choice of profession after being on board three sinking ships. However, she continued to work for cruise and shipping companies throughout her career. Despite an amazing level of coincidence, Violet Jessop had no bearing on the three events. She merely had a hapless employment history. Violet did not cause a single shipwreck, let alone all three.

---

[8]"Violet Jessop." Biography.com. April 02, 2014. Accessed June 09, 2018. https://www. biography.com/people/violet-jessop-283646.
[9]"HMS Hawke (1891)." Wikipedia. June 02, 2018. Accessed June 09, 2018. https:// en.wikipedia.org/wiki/HMS_Hawke_(1891).

Our adventures in quantitative research are not nearly as hazardous, but we do witness coincidences on a regular basis. Sales briefly increase. Page views temporarily decline. "Likes" momentarily stagnate. These behaviors are noticeable, but are they notable? Coincidences experienced during a project can often mislead us into making reactionary and shortsighted decisions. We will discuss three common hazards of research that lurk beneath the surface of your projects and sink good ideas.

Texas Sharpshooter Fallacy

Imagine for a moment a gritty cowboy, the type of fella that might have a mouthful of chewing tobacco and hips flanked by two Colt .45 revolvers. A western sun offsets his dusty silhouette, as tumbleweeds blow by in the distance. Our cowboy stands motionless, guns at the ready, staring with focused attention at an old, wooden barn standing several yards away. He spits, raises his revolvers, and quickly fires 12 shots.

As the dust clears, we see bullet holes scattered across the barn's wooden wall in no apparent order or pattern. A few shots hit near the center of the wall. Some hit near the roof. Others hit near the foundation. The cowboy walks up to the barn, pulls out a piece of chalk from his pocket, and draws a single, continuous line around all the bullet holes. His drawing forms a large, weirdly shaped outline. Upon its completion, the cowboy exclaims, "Well, look'y here. All my shots hit the target!"

We can all be Texas sharpshooters if we do not carefully evaluate the entirety of the available data. Simply looking for clusters that align with our biases may lead us to incorrect conclusions.

For example, the review of a website's analytic information serves as an excellent resource to evaluate past performance. However, we can use analytics to predict future performance only if the website stays the same, devoid of any design or technical changes. To do otherwise would be like trying to count old bullet holes in a new barn. Once you introduce changes to an experience, analytic information becomes purely historical. Until you accumulate a sufficient mass of new information, analytics are irrelevant. New barns only show new bullet holes. Even then, you still might draw the wrong target.

Draw your target before you evaluate data. This sounds simple, but even experienced pros sometimes misunderstand this concept. Consider the following scenario:

> Acme Company changes their website's home page and wants to evaluate its aesthetic merits. They measure the number of visits. After making the change to the home page, fewer visitors view the page. Therefore, Acme Company believes the new design is less successful than the previous one.

In this example, Acme Company counts bullets (the number of visits) on the target (the site's home page). Outside of search engine optimization, the number of visits rarely has anything to do with a page's visual design. After all, a visitor could view the page and say, "I think this home page looks horrible," and then leave. However, analytics software still counts his or her visit. A page visit is an ineffective means of evaluating visual design. The number of visits reflect market awareness and supporting media efforts, but not the page's visual design. Acme counted bullets but chose the wrong target.

Paint your target, then count the bullet holes. You will be a sharpshooter in no time.

Procrustean Bed

If the Texas sharpshooter fallacy exuded a certain country charm, the story of the Procrustean bed should scare the hell out of you. According to Greek mythology,[10] an old ironsmith named Procrustean would offer shelter to weary travelers along the road to Athens. While they slept, Procrustean would strap the travelers to their beds and stretch their bodies to fit the bed frame. Short people got off easy. The tall ones truly suffered. Procrustean chopped off their feet, ankles, and shins until the travelers fit neatly into their beds.

You find Procrustean solutions frequently in quantitative research. Data is stretched and truncated to meet a chosen outcome. Business objectives are overplayed; user needs are downplayed. Device requirements are overplayed; affordability is downplayed. Gesture controls are overplayed; the aging population is downplayed. Stretch. Chop. Enhance. Remove. We become data sadists.

We also affect data while collecting it. Selection bias stretches and pulls data by altering whom or what we select as the data's source. Research trends, such as "get out of the building" (GOOB), can be a powerful tool to solicit feedback from users. Here, we leave our offices and visit a public setting. We find users and show them an app or website, engaging and testing how the audience responds. However, like Procrustean sizing up his guests on the road to Athens, we may inadvertently—or intentionally—select users based

---

[10]"Procrustes." Princeton University. Accessed May 28, 2018. https://www.princeton. edu/~achaney/tmve/wiki100k/docs/Procrustes.html.

on non-representative criteria. We subconsciously select people who look friendly, relaxed, and outgoing. On-the-street interviews, retail intercepts, and all face-to-face interactions carry the possibility that we may reach only those people who are willing to talk to us. Are they representative of your audience, or are they only representative of people willing to talk to an inquisitive stranger holding an iPad?

Keep a vigilant eye on data that fits a little too neatly into recommendations—even your own. Realistic assessment of data may occasionally clip your wings, but it will help you avoid getting cut off at the knees.

Hobson's Choice

Livery stables were the 17th-century equivalent to today's car rental companies. Riders chose a horse, rode it, and then returned it. Thomas Hobson[11] ran a livery stable outside of Cambridge, England. He realized that riders chose the good horses far more often than the bad, resulting in the overuse of some horses and the underuse of others. Like automobiles, horses accrue mileage. Hobson decided to eliminate the rider's choice. He gave prospective riders a single option: ride the horse I choose for you or do not ride at all. In short, "take it or leave it."

We often face a Hobson's choice when researching, designing, and building software. We accept a bad solution rather than go without. A study does not include enough participants; an experience feels awkward; an app's performance trots rather than gallops. However, your team employs the solution anyway. Short schedules and insufficient budgets often take the blame.

If a solution were bad, it would be best to not take it out of the stable, so to speak. In today's world of rapid iteration, we sometimes accept a Hobson's choice solution in the hope that eventually it will be replaced. We emphasize the now over the good at our peril. As the saying goes: "The joy of an early release lasts but a short time. The bitterness of an unusable system can last for years."[12]

Researching, brainstorming, designing, developing, scheduling, budgeting, and managing generates a lot of horse shit. You need to find a way to stomp through it and reach the road leading to your audience. Avoid the hazards along the way. Recognize coincidence, pick your targets, and always be wary of strange, old men offering help—including me.

---

[11]"Thomas Hobson." Wikipedia. June 08, 2018. Accessed June 09, 2018. https://en.wikipedia.org/wiki/Thomas_Hobson.
[12]Pavelin, K., J.A. Cham, P. De, C. Brooksbank, G. Cameron, and C. Steinbeck. "Bioinformatics Meets User-centred Design: A Perspective." *Advances in Pediatrics*. July 12, 2012. Accessed June 09, 2018. doi:10.1371/journal.pcbi.1002554.

Key Takeaways

- Quantitative research involves numerical and statistical analyses.

- Quantitative research provides the "what" about a phenomenon.

- Useful quantitative research is statistically significant.

- Good statistics are generalizable and can be used to infer conclusions about an entire population.

- When collecting data, make sure research subjects are representative of a population.

- Correlation is not causality!

Questions to Ask Yourself

- If the research were repeated, how often would it produce similar results?

- How accurate is the research data?

- Is our research data representative of a population?

- Am I inadvertently selecting people from a population who are like me?

- Am I accounting for people who do not respond to a survey?

- Am I mistaking a correlation for a causality?

- Have I clearly stated my research objective before conducting research?

- Am I ignoring differences or overemphasizing similarities within the research data?

- Am I stretching data to meet my client's, my team's, or my own needs?

- Is a timeline or budget affecting my objectivity?

Calculator Research

In late 2013, McResources, McDonald's employee website, posted curious advice[1] to its hourly workers: it suggested the appropriate amount to tip their au pairs, pool boys, and personal trainers.

For context, the Bureau of Labor Statistics reports food preparers, such as McDonald's employees, earn a median pay of \$10.93 per hour[2] (see Figure 34-1); childcare workers (au pairs), \$10.72;[3] grounds maintenance workers (pool boys), \$13.51;[4] and fitness (personal) trainers, \$18.85.[5]

---

[1]Calabrese, Erin, and Josh Saul. "That's Rich! McDonald's Tells Workers What to Tip Au Pairs." *New York Post.* December 07, 2013. Accessed June 09, 2018. https://nypost.com/2013/12/07/rich-with-irony-mcdonalds-gives-workers-advice-on-tipping-pool-boys-au-pairs/.
[2]"Summary." U.S. Bureau of Labor Statistics. Accessed June 4, 2018. https://www.bls.gov/ooh/food-preparation-and-serving/food-preparation-workers.htm.
[3]"Summary." U.S. Bureau of Labor Statistics. Accessed June 4, 2018. https://www.bls.gov/ooh/personal-care-and-service/childcare-workers.htm.
[4]"Summary." U.S. Bureau of Labor Statistics. Accessed June 4, 2018. https://www.bls.gov/ooh/building-and-grounds-cleaning/grounds-maintenance-workers.htm.
[5]"Summary." U.S. Bureau of Labor Statistics. Accessed June 4, 2018. https://www.bls.gov/ooh/personal-care-and-service/fitness-trainers-and-instructors.htm.

© Edward Stull 2018
E. Stull, *UX Fundamentals for Non-UX Professionals*,
https://doi.org/10.1007/978-1-4842-3811-0_34

Figure 34-1. Bureau of Labor Statistics website showing median pay of food preparation workers[6]

A full-time, $10.93-per-hour McDonald's worker would earn approximately $22,730 per year (2,080 hours). Assuming this worker hired an au pair to cover her full-time schedule, she would spend $22,290 per year on childcare alone. If the McDonald's worker had her pool cleaned for an hour once a week, she might expect to pay a meager $702.52. Assuming she can make it to the gym, too, she would pay an additional $980.20 for her personal trainer. This leaves the McDonald's worker $1,242.72 in debt annually before taxes, before all other living expenses—and before tipping. Not a lot of Happy Meals.

---

[6]"Food Preparation Workers." Digital image. U.S. Bureau of Labor Statistics. Accessed June 7, 2018. https://www.bls.gov/ooh/food-preparation-and-serving/food-preparation-workers.htm.

Clearly, McDonald's hourly workers do not hire many au pairs, pool boys, and personal trainers. The press pilloried the company for its tone-deafness to the economic challenges of its employees. The mistake was regrettable and avoidable.

Quantitative research does not always demand an extensive research study. Sometimes, it simply requires a calculator.

Key Takeaways

- Calculate known values to assess the plausibility of quantitative research data.

Questions to Ask Yourself

- Does a numerically based claim sound reasonable?
- What is the average age and income of a population?

Qualitative Research

In 2013, Nike released a women's fashion line based on Pacific Islander tattoos. The line consisted of jump suits, singlets, and sports bras. Each item displayed attractive, ornate designs referencing traditional Samoan patterns. Each design attempted to pay homage to its source, but a lack of research turned a cultural tribute into a media nightmare.[1]

By all accounts, traditional forms of tattooing[2] (tatua) are both extremely painful and occasionally dangerous. Polynesian artists apply delicate line work and bold triangular blocks of ink with combs fashioned from fish bone, turtle shell, and boar tusk. Razor-sharp, ink-laden combs drive through the recipient's skin with the force of an artist's mallet. Blood flows. Sunset grants a nightly reprieve to the process that may last for several days. Compared with Western forms of tattooing, these indelible patterns are earned only through suffering. Both men and women wear traditional tattoos, each gender displaying a distinctive style: pe'a[3] for men, malu[4] for women (see Figure 35-1).

[1]"Nike Pulls Tattoo Leggings After Offending Pacific Community." The Huffington Post. August 15, 2013. Accessed June 09, 2018. https://www.huffingtonpost. com/2013/08/15/nike-tattoo-leggings_n_3763591.html.
[2]"Polynesian Tattoo: History, Meanings and Traditional Designs." Zealand Tattoo. Accessed June09,2018.http://www.zealandtattoo.co.nz/tattoo-styles/polynesian-tattoo-history-meanings-traditional-designs/.
[3]"Pe'a." Wikipedia. February 17, 2018. Accessed June 09, 2018. https://en.wikipedia. org/wiki/Pe%27a.
[4]"Malu." Wikipedia. April 03, 2018. Accessed June 09, 2018. https://en.wikipedia.org/ wiki/Malu.

© Edward Stull 2018
E. Stull, *UX Fundamentals for Non-UX Professionals*,
https://doi.org/10.1007/978-1-4842-3811-0_35

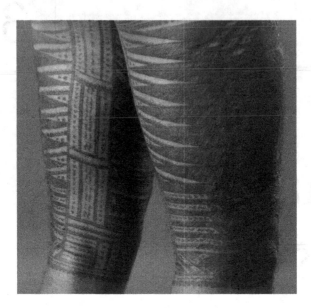

Figure 35-1.. Man with traditional pe'a (tattoo)[5]

Nike experienced its own brand of discomfort with pe'a and malu. When the company designed its women's tech gear, it chose a gorgeous arrangement of one of the two traditional styles. The problem for Nike was that it chose the wrong one. It placed the men's pattern on the women's clothing. A simple choice created a complex problem. Though Nike's intentions were perhaps good, the choice angered Pacific Islanders. The cultural faux pas generated condemnations of Nike's cultural insensitivity. Within weeks, Nike had pulled the entire line from store shelves and retreated with a heartfelt apology.

Groups of people tend to share similar patterns. Polynesians may originate from the same island communities, adorn themselves with the same traditional tattoos, and undergo the same painful processes. Some patterns delight whereas others offend. We discover these patterns through research.

Every population has its own patterns. Tipping in United States. Eating with your right hand in Oman. Refusing a gift three times in China. Patterns unite people.

Yet, each person is an individual, experiencing the world in his or her own unique way. When we study individuals, our view is framed by their lives alone. You may enjoy vacations at a high-priced Hawaiian resort, drink Cristal champagne, and quote Nietzsche. Your neighbor may relax on his camo-patterned La-Z-Boy recliner, swig Coors Light, and read trashy romance novels. We study the qualities of such data—the qualitative.

---

[5]Cropped, Photo by CloudSurfer, "Traditional Samoan Tattoo," 2002.

Because we study a small population of people, the resulting data may not represent anything more than the individuals we study. You may enjoy vacations at high-priced Hawaiian resorts, but that does not mean everyone sharing your gender, age, nationality, income, or profession does. No, instead, qualitative research dives deep into an individual's culture, history, and behaviors. We discover questions we had never thought to ask.

Discovering Questions

Qualitative research uncovers the culture, history, and behaviors of a population. Culture need not be that of an entire society; it may be limited to a profession, organization, or family. History need not cover an entire era; it may be restricted to a few months, weeks, or days. Behavior need not include all activity; it may simply be a workflow, function, or gesture.

Consider hospital nurses. Imagine the life of a nurse working in Brookville, New York. Brookville has one of the highest average net worth of any town in America, averaging around 1.8 million dollars.[6] Fewer than 900 households fill its four square miles. The town served as inspiration for the fictional town of West Egg, the setting of F. Scott Fitzgerald's book, *The Great Gatsby*. What type of cases does a nurse in Brookville treat? Is she witnessing premature births by teen mothers, or the ravages of methamphetamine abuse? Not likely.

Now, imagine the life of a nurse working in Allen, South Dakota. With a median household income of less than $14,000,[7] the citizens of Allen face hardships many of us would find difficult to imagine. A thousand miles from the nearest coastline, the town sits within the Pine Ridge Indian Reservation. It has the highest poverty rate in the United States. What type of challenges does a nurse in Allen contend with during her rotation? Not the same as those experienced by a nurse in Brookville.

To understand a population, we could further explore statistics such as marital status, educational attainment, and social media use. You can find a wealth of such data on the web, including the U.S. Census, the Bureau of Labor Statistics, and Pew Research Center. However, statistics only imply what it is like to be a nurse. It tells us nothing about nurses' daily lives as they experience the joys of birth, the sorrows of death, and the drudgeries of paperwork—along with the many winks, tears, and eye-rolls.

---

[6]"Brookville, NY." Data USA. Accessed June 09, 2018. https://datausa.io/profile/geo/brookville-ny/.
[7]"Allen, SD." Data USA. Accessed June 09, 2018. https://datausa.io/profile/geo/allen-sd/#heritage.

Groups of people are rarely homogenous, but they do tend to share at least a few commonalities. Nurses may share similar joys and struggles. Tattooed Polynesians may share the same traditional patterns. But to fully understand people, we must observe.

Contextual Inquiry

In her book, *To Kill a Mockingbird*, Harper Lee wrote: "You never really understand a person until you consider things from his point of view... Until you climb inside of his skin and walk around in it."

I imagine you reading this book. As most people tend to sit while reading, I picture you sitting in a chair. As most people tend to read in a well-lit room, I picture you reading next to a lamp. As most people tend to live in a family household,[8] I picture you living with others. As most people tend to own a pet,[9] I picture you reading next to a dog or cat.

Statistically speaking, my assumptions are defendable. The majority of American adults reading a book likely do so while sitting in lit rooms in proximity to both their families and pets. Though this information may illuminate some of your attributes, it certainly does not tell me much about your life. You are not a pie chart.

Perhaps you are reading this book while running on a treadmill. Perhaps you are reading this book on an e-reader sitting in a dark room. Perhaps you are reading this book to relax after taking care of an aging parent. Perhaps you are reading this book as your helper monkey gets you a beer. These qualities may have a statistical reference: we could find the number of households with pet spider monkeys, aging parents, e-readers, and gym memberships. The question is: would you even think of doing so, unless you observed it in person?

Contextual inquiry is an ethnographic method by which we observe and interview people within their own environments: homes, offices, coffee houses, churches, soccer fields, plane cockpits, and the like. We witness their joys and pains. Through this observance, we discover the hidden attributes and behaviors of an audience: qualities that we would not even think of researching until we observed them firsthand.

---

[8]Lofquist, Daphne, Terry Lugaila, Martin O'Connell, and Sarah Feliz. "Households and Families: 2010." Households and Families: 2010. April 01, 2012. Accessed June 22, 2018. https://www.census.gov/library/publications/2012/dec/c2010br-14.html.
[9]"Pet Industry Market Size & Ownership Statistics." 2017-2018 APPA National Pet Owners Survey. Accessed June 22, 2018. http://www.americanpetproducts.org/press_industrytrends.asp.

Years ago, I helped a client design a customer call center application. The application provided customer service representatives (CSRs) a means to quickly retrieve product information. The CSRs worked within a huge hangar-sized facility. The building's high ceilings reached 30 feet at the center and sloped to exterior walls dotted with vending stations and small conference rooms. Rows of waist-high cubicles ran from one side of the massive building to the other like long lines of dominos. Each cubicle—no more than a few feet wide—contained a monitor, computer, keyboard, and a headset. A central computer routed calls to available CSRs. Upon receiving a call, a CSR would walk a customer through a series of scripted questions leading to a product offer.

A CSR would read aloud from his or her computer screen based on the customer's responses to the scripted questions. "Yes" answers directed the script along one path; "no" answers diverted the script to another.

As you might imagine, sitting for hours, taking dozens of calls, and reading from a computer screen leads to strained eyes, sore legs, and aching backs. It is mind-numbingly boring as well. The CSRs are required to read from an approved on-screen script, click buttons, and type customer responses into form fields. After a few hours, even the calmest of individuals would become fidgety. CSRs squirm in their chairs and slide away from their desks—the same desks on which their computer screens sit. To appreciate this behavior, please read the following aloud:

> Hi, thanks for reading this book. I appreciate it.

Now, this is important: please hold this book in your right hand, stretch your arm out as far as you can, and read the following line aloud:

> Hi, again. Reading from this distance is hard. Isn't it?

I did not realize that CSRs moved away from their computer screens until I saw it myself. After a few hours in a chair, people would put up their feet on their desks. Nobody mentioned it during related surveys or interviews. Doing so was natural and unremarkable. The behavior had to be observed.

In this particular case, the observation that CSRs slowly move away from their computer screens led to a considerable increase of on-screen text sizes. Rather than struggle to read the text, CSRs could now move to any position they wished. They put up their feet, sat back, and read at their leisure. I hope you are doing the same.

Interviews

Inspector Clouseau: Does your dog bite?

Hotel attendant: No.

(Dog then bites Inspector Clouseau.)

Inspector Clouseau: I thought you said your dog did not bite!

Hotel attendant: That is not my dog.

Interviews are funny. Not necessarily funny in the same way as this quote from the 1976 film *The Pink Panther Strikes Again*,[10] but funny nonetheless. They are unwieldy exchanges between two people, full of potential insights, surprises, biases, and fears.

"What do you think of ACME's website?" we ask. The interviewee replies, "It's fantastic!" If we were to end there, we might assume the website is perfect. Project complete. But, if we were to follow up with, "Is there anything you would change with this application? If so, what would it be?" an interviewee might say anything from, "Yes, I'd change this period to a comma" to "Yes, I'd change your entire business model." A follow-up question is worth a dozen one-offs. People are wonderfully unpredictable, as they go from rational beliefs to surprising absurdities.

All interviews begin with a single question. Yet, it can be almost anything. How does one become a police inspector? What was it like to steal the largest diamond in the world? Whether researching products or interrogating suspects, a question is your most powerful research tool. However, for a question to be effective, it must be open-ended and dispassionate.

Open-Ended Questions

If I were to ask you where you were born, you would likely reply with a city name. However, if I asked you to *tell me about where you were born*, you might say a lot more. The first question was closed-ended, the second was open-ended.

Closed-ended:

> "Where were you born?"

> "Orgelet."

---

[10] *The Pink Panther Strikes Again*. Directed by Blake Edwards. United States: United Artists, 1976.

Open-ended:

> "Tell me about where you were born."

> "Orgelet is a town on the western coast of France. We have a marathon each year. Lots of drunk people show up for it."

Rather than ask yes/no questions, pose open-ended questions that start with "Tell me," "Describe," or "Explain."

Leading Questions

When you care about something, unchecked biases can slip into your work. Asking questions is no different; the hand of the author sometimes shows. To yield reliable answers, we must avoid leading a respondent to a particular response. A typical leading question reads like the following:

> Do you think this website performs poorly?

You may feel that a website performs poorly, but asking such a question prejudices a response. The phrasing would lead the respondent to think the website is performing poorly, potentially skewing her or his answer toward the negative. Likewise, if you ask whether the website performs well, it may skew an answer toward the positive.

A non-leading alternative phrasing of the question is the following:

> What are your thoughts about this website?

This question removes our bias. The question does not reveal our impressions of the website. Respondents reply with their answers, not yours.

Leading questions are often far less obvious than our previous example. They may be accidental and asked in good faith. Consider the following:

> How would you improve this application?

At first read, the question is innocuous: we are simply soliciting a respondent's opinion. We are not asking if the application is good or bad. But, here too we subtly influence the response. Does a good application need to be improved? By asking what a respondent would improve, we have implied that something could be improved; that the application is lacking in some way. Again, we can pose a non-leading alternative:

> Is there anything you would change with this application? If so, what would it be?

By rephrasing "improve" to "change," we eliminate the implied judgment contained within the question. However, we still imply that something may need to be changed. This is usually okay, as the follow-up question gives the respondent a way out: he or she can reply "No" to the first question, thereby skipping the second.

Loaded Questions

Coercive phrases may affect answers. It is sobering to realize how easily we can be manipulated. Consider the following:

> What activities to you enjoy doing most while using a
> tablet device?

Respondents may not enjoy using a tablet device at all. Perhaps they do not even own one. Our biases may steer us to believe that most people own and enjoy using tablets, but you need not look far beyond your own socioeconomic bubble to find exceptions. Ask someone making less than $30K per year and living in a retirement home. According to Pew Research Center, tablet ownership tends to dip considerably in such populations.[11]

Interviewees are often humble, occasionally entertaining, sometimes snarky, and rarely hostile. (But it happens.) We must tread careful when it comes to the language we use. Emotionally charged terminology skews responses and risks turning an interview into a debate. One person's euphemism is another person's insult. Pro-life. Freedom fighter. Sissy. Illegal immigrant. Regime change. Flyover state. Secretary. Victim. Crippled. Crazy. Senile. Reject. Junkie. Did your heart race upon hearing a few of these? So too will your interviewees' if you use emotionally charged language. Best to keep things conversational, or learn to use a defibrillator.

Silence

Interviewees are protagonists in their own stories, and their stories are told in the first person. The only knowledge they possess is their own. Interviewees are not omniscient. They want to appear helpful and smart, and sometimes to avoid looking stupid, they stop talking.

Dead air. Pregnant pauses. Non sequiturs. Although such moments may feel awkward, silences lead to goldmines of information. Pauses allow interviewees to collect their thoughts, even if for only a few seconds. Out of courtesy, interviewers may be tempted to fill in moments of quiet. However, we should

---

[11]Anderson, Monica, and Andrew Perrin. "1. Technology Use among Seniors." Pew Research Center: Internet, Science & Tech. May 17, 2017. Accessed June 22, 2018. http://www.pewinternet.org/2017/05/17/technology-use-among-seniors/.

wait for a response. As interviewees grow more comfortable with our questions, the pauses in between allow them to digress into other topics. A question about paying a hotel bill leads to answers about room service. A question about dog food leads to answers about leash laws. Whatever the topic, an interviewee's potential insights are magnitudes greater than any list of questions we might prepare. We ask questions not only to elicit answers, but to also uncover questions we never thought to ask.

False Data

On average, people lie three times during a 10-minute conversation.[12] Our deceptions are usually unconscious: we skew our answers to make ourselves look good in someone else's eyes, as well as in our own. If you ask an interviewee, "How many times a week do you exercise?" You will receive an answer, but it may be more aspirational than factual. Asking a person how many times they exercise per week is a direct question. Direct questions are quick but the answers may be unreliable.

The alternative to asking a direct question is, unsurprisingly, to ask an indirect question. Using our previous example, we wish to uncover the number of times the interviewee exercises per week. Rather than ask the question directly, we instead say to the interviewee, "Please explain your typical week." The interviewee tells us she exercises after dropping of her kids for their piano practice on Tuesdays. She adds, "I try to exercise every Saturday, too." In conclusion, the interview indicated that she exercises twice a week.

Occasionally an interviewee may anticipate your line of inquiry and give intentionally false information. For example, asking employees about how their accounting software performs may imply their manager is debating changing it. In an effort to support the presumed change, an employee might skew his evaluation of the software's performance downward. You could watch for the telltale signs of deception, such as maintaining too much eye contact, sitting in a frozen posture, and changing contractions into two separate words, or you could simply ask an indirect question, as follows:

You ask, "How do you calculate profit and loss?"

"Oh, I hit this button here" he replies.

"Is there anything you would change about this process?" you ask.

"Hmm... I don't think it could be improved—it's super easy" he replies.

---

[12]"UMass Amherst Researcher Finds Most People Lie in Everyday Conversation." Freedom National Bank Opening. June 10, 2002. Accessed June 09, 2018. https://www.umass.edu/newsoffice/article/umass-amherst-researcher-finds-most-people-lie-everyday-conversation.

Asking indirect questions is more time-consuming than direct questions, but you will find the answers far more straightforward.

Group Interviews

When interviewing multiple people, you indirectly receive a wealth of information by asking one interviewee about another. This technique works in both individual and group interview settings. In individual interviews, your last question might be, "I'm meeting with Sally Salesperson after you. Anything I should ask her?" I have learned that interviewees were getting married soon, resigning the next day, skeptical of their upcoming interviews (always good to know in advance), and have been warned that an interviewee had "the gift of the gab." In group interviews, posing questions about fellow interviewees can be a good conversational accelerant. "Bob, you send Steve customer order information. How does that work?" Having Steve in the room to hear Bob's response may lead to agreements or challenges. Both are good. But keep in mind that some participates will dominate a group discussion. Try to interview participants individually as well.

The Five Whys

The next time you interview someone, try the "five whys." Developed by Toyota,[13] the technique uncovers answers through the repetition of a single question: why? At first, "the five whys" can sound like the nagging of a precocious child, "Why? Why? Why? Why? Why?" Imagine the following interview situation:

1. You: Why do you want to create an iPhone app for hotel attendants?

    Interviewee: Because hotel attendants need a quick way to summon a bellhop.

2. You: Why summon a bellhop?

    Interviewee: Because bellhops carry guests' bags to their rooms.

3. You: Why do bellhops carry guests' bags to their rooms?

    Interviewee: Because we want to show courtesy to our guests.

---

[13] Ohno, Taiichi. "Toyota Global Site | Ask 'why' Five times about Every Matter." TOYOTA MOTOR CORPORATION GLOBAL WEBSITE. March 2006. Accessed June 09, 2018. http://www.toyota-global.com/company/toyota_traditions/quality/mar_apr_2006.html.

4. You: Why show courtesy to your guests?

   Interviewee: Because courtesy is this hotel's unique offering.

5. You: Why is courtesy the hotel's unique offering?

   Interviewee: Because we are more expensive than our competition, so we compete by providing better service.

Starting with one simple "why," we were able to generate four additional questions, eventually getting to the heart of an issue. Could we have discovered this information another way? Certainly. However, the "five whys" uncover an interviewee's underlying assumptions and motivations. Interviewees will freely tell you what they believe. We must discover *why* they believe it. Without such understanding, we can only guess the reasons.

Key Takeaways

- Qualitative research provides the "why" about a phenomenon.

- Small sample sizes may not be representative of an entire population.

- We run contextual inquiries to observe people within their own environments.

- Effective qualitative research questions are open-ended and dispassionate.

- Open-ended questions start with phrases such as "Tell me," "Describe," or "Explain."

- Leading questions prejudice a response and compromise research efforts.

- Coercive questions and emotionally charged language skew responses and risk turning an interview into a debate.

- Silent pauses allow interviewees to collect their thoughts and digress into additional topics.

- The "five whys" technique uncovers answers through the repetition of a single question: why?

Questions to Ask Yourself

- What patterns are shared amongst a population?

- How can I change a closed-ended question into an open-ended one?

- How can I observe participants in their own environments?

- Am I asking participants any leading questions?

- Do the questions I ask participants contain opinions or value judgments?

- Am I asking questions in a dispassionate or coercive way?

- Do any of my questions contain emotionally charged language?

- Am I adequately pausing after each question?

- Am I inadvertently filling in silences during an interview?

- Does an interview contain a mix of direct and indirect questions?

- What can I learn about research participants from one another?

Reconciliation

From the rural towns of Mexico to the hillside villages of Honduras lives the Xoloitzcuintli[1] (see Figure 36-1). This unusual breed of dog traces its lineage farther back than the Aztec Empire. Although rare, you can still find the Xolo today. The American Kennel Club estimates their numbers to be nearly 30,000 worldwide: a mere whimper compared to the deafening howl of millions of Labrador Retrievers. With its hairless coat, huge ears, and occasional Mohawk, the Xolo is a frequent contestant in ugliest dog competitions. Despite these momentary humiliations, the breed is revered for its gentle personality and a somewhat surprising mythology—it has magical healing powers. Some people say the same about UX.

Figure 36-1. Artist's rendering of a Xoloitzcuintli

---

[1] "Xoloitzcuintli Dog Breed Information." American Kennel Club. Accessed June 09, 2018. https://www.akc.org/dog-breeds/xoloitzcuintli/.

© Edward Stull 2018
E. Stull, *UX Fundamentals for Non-UX Professionals*,
https://doi.org/10.1007/978-1-4842-3811-0_36

Cuddling up with the Xolo is rumored to help with everything from asthma to toothaches. A few people even attribute the dog's healing powers to its ability to ward off evil spirits. But as with all such mythology, the help it provides may be real, but the reason why remains a mystery. The myth likely stems from the Xolo's warmth. The breed runs hot. Combined with its hairless coat, the dog becomes a portable little heater with a Mohawk. One can imagine the Xolo soothing its owner during times of illness, providing comfort for a range of ailments.

UX is the Xolo of the digital world. Compared to the allure of visual design, UX can be as unsightly as a contestant in an ugly dog competition. Its outputs look unsophisticated. Its research appears unwieldy. And, like the Xolo, myths abound. UX is reputed to help with everything from failed marketing strategy to poor project planning. But, in reality, UX can improve digital products through one action alone: the hard work of reconciling information.

The primary cause of any UX problem is the accidental or intentional avoidance of reconciling information. Perceptions must be clarified, and contradictions must be settled.

Most of what a UX researcher does consists of discovering disparate ideas, thoughts, and opinions. She or he then reconciles this information into a cohesive set of findings. Surely, several variations of this work exist; yet, all UX research involves some form of reconciliation.

If you asked a business what a perfect app might be, you would perhaps get answers involving what users give the business (e.g., money, time, ad impressions, etc.). If you asked users what a perfect app might be, you would perhaps get answers involving what the business gives users (e.g., utility, entertainment, etc.). This is the fundamental dichotomy between user needs and business goals: some experiences favor the business, some favor the user, and some are mutually beneficial.

The careful reconciliation of user needs and business goals is where great software is born. Build an app that addresses only business goals and no users will use it; build an app that addresses only user goals and you'll likely go out of business. For example, consider the following two apps:

Acme "Give Us $1" app:

> One screen with one button labeled: "Give Us $1"
>
> Upon tapping the button, the user gives Acme one dollar.

Acme "Get Your $1" app:

> One screen with one button labeled: "Get Your $1"
>
> Upon tapping the button, the user receives one dollar from Acme.

The Acme "Give Us $1" app is a shining example of where a business can go wrong with user experience. The business gives nothing to the user in return for his or her dollar. Getting one dollar for doing nothing would be welcomed by many businesses; however, the number of users willing to participate would be extremely limited.

The second example, the Acme "Get Your $1" app, exemplifies a business only addressing user needs. The user receives a dollar for doing nothing. This business will not be in business for long.

Granted, these examples are simplified to show the extremes of user needs and business goals. From banking websites to gaming apps, these exchanges happen billions of times of day. The good applications exchange value, the bad ones do not.

Think about the applications you use daily. A banking website allows you to pay bills online. In return, you reciprocate by allowing the bank to hold your deposits. Facebook gives you the ability to like and post messages. In return, those behaviors build Facebook's inventory of marketing data. The exchange could be purely monetary, such as e-commerce, or it could be a matter of exchanging money for entertainment, such as a video game or movie rental.

The basis for design is the reconciliation of user goals and business objectives. Everything is possible, but needs and goals either complement or compete with one another. Knowing which one does what is where design begins.

Do It Now or Do It Later

Some of the best UX designers I know would never call themselves UX designers at all—they would call themselves front-end developers. The practical reality is that developers make UX design decisions on the fly all the time, sometimes in the late-night hours before a product launch. Why do developers get pressed into the role of UX designer? The reason is simple: UX design decisions are unavoidable; you either make these decisions before development or during it.

Shortsightedness is not limited to any particular role or activity, but it compromises UX research efforts the most. What problem are we solving for users? How did they get here? What happens if there is an error? Such questions and countless more are present within your project. The questions may seem minor today, but the answers can come back to bite you.

Roy Peter Clark, author of *Writing Tools: 50 Essential Strategies for Every Writer*[2], states that to truly understand a subject you need to "get the name of the dog."

---

[2]Clark, Roy Peter. *Writing Tools - 50 Essential Strategies for Every Writer.* Little, Brown & Company, 2008.

Journalists learn a lot when visiting the scene of events. They see everything from buildings on fire, to riot police in full gear, to jumpers on window ledges. However, while this information is vital to the news story, it is not the detail that makes the story come to life for the reader. A traumatized family standing on the street watching their house burn is a tragedy, but knowing the name of the dog that sits beside them attentively is what makes it a story. To achieve this level of connection with an audience, you need to do the work: you need to get the name of the dog.

While this reference pertains to journalism, it is applicable to UX research. Along with uncovering small details, you discover major issues in the process. For example, consider the following:

> Small detail: What if a user is color blind?
>
> Major issue: Does your product need to be ADA compliant?
>
> Small detail: Will people use your website at work?
>
> Major issue: Do employers block your website?
>
> Small detail: Do you need to display an EU privacy policy?
>
> Major issue: What happens if a user rejects it?

Minor details shape a well-known story about the rock band Van Halen. (Coincidentally, the band's former front man, David Lee Roth, is an avid dog trainer.) As told in Roth's autobiography, *Crazy From the Heat*[3], Van Halen's contracts with tour promoters included a few unusual instructions. In the midst of complicated language about the specific electrical configurations, logistical support, and other contractual clauses associated with putting on a rock show was the line: "There will be no brown M&Ms in the backstage area, upon pain of forfeiture of the show, with full compensation." The line read as an eccentric request but it served an important purpose: if the tour promoters skipped a minor detail with such severe consequences, the band would know the promoter likely skipped over more important issues, such as the electric configurations and logistical support they required. The pursuit of minor details frequently leads to uncovering major issues.

Users rarely provide such clear instructions. Some make requests. Some complain. More often, they simply abandon. And, when they do, they take their minor details and major issues with them. Likewise, businesses rarely know what to ask for during the development of a digital project. Complaints come too late. Rather than evolve an existing product or service, many companies scrap it and start over—yet another form of abandonment. However, we can

---

[3]Roth, David Lee. *Crazy from the Heat.* London: Ebury, 2000.

help stave off this unpleasantness by doing the hard work now and asking tough questions. Does an experience ask too much of a user? Does it ask too much of the business? Is value exchanged? You might be surprised by what you uncover. Although research does not guarantee a successful experience, it often keeps you from barking up the wrong tree.

Key Takeaways

- The primary cause of any UX problem is the accidental or intentional avoidance of reconciling information.

- All research involves some form of reconciliation.

- The basis for design is the reconciliation of user goals and business objectives.

- UX design decisions are unavoidable.

- The pursuit of minor details frequently leads to uncovering major issues.

Questions to Ask Yourself

- Am I avoiding any information?

- What contradictory, difficult, and unpleasant information exists about a topic?

- How does each business objective align with a user goal?

- What's the name of the dog?

Documentation

For millennia, skilled carvers have chiseled stone into commandments, declarations, and epic tales of triumph and struggle. Skilled as these craftsmen were, their chosen medium had a disadvantage: its weight. At over 1,600 pounds[1], it was no surprise the Rosetta Stone sat sedentary for 2,000 years before anyone dreamed of digging it up and giving it a read.

Writers welcomed the invention of lighter writing materials, scribbling on everything from birch bark to beeswax. Hammered animal hides and sunbaked plants gave way to pulped wood and rag fibers. Each evolution became more portable. Spreading across the ancient world, paper would come to hold the foundational stories of religions, philosophies, and the sciences. Yet, many of these stories did not endure. Without our attention, history disappears.

Though more mobile than a stone slab, paper was far less durable. Insects ravaged papyrus and parchment, often leaving only traces of the past. But fire proved to be history's true enemy, sparing few documents from lightning strikes, misplaced candles, and fanatical firebugs.

In 221 BCE, China's first emperor ordered all history books to be set ablaze[2], only safeguarding his imperial archives. Years later, in retribution, rebelling troops burned those as well. Around the same time, the famed Library of Alexandria was rumored to have been burned. However, some of its works may have survived in the Serapeum, a nearby temple. Years later, that too,

---

[1]"Rosetta Stone." BYU-Idaho. Accessed June 09, 2018. http://www.byui.edu/special-collections/exhibits/rosetta-stone.
[2]Britannica, The Editors of Encyclopaedia. "Mao Chang." Encyclopædia Britannica. July 20, 1998. Accessed June 09, 2018. https://www.britannica.com/biography/Mao-Chang#ref846671.

© Edward Stull 2018
E. Stull, *UX Fundamentals for Non-UX Professionals*,
https://doi.org/10.1007/978-1-4842-3811-0_37

burned. Its surviving documents may have been relocated to the far-off Imperial Library of Constantinople.[3] And, if you have not guessed what happens next, in 1204, knights from the Fourth Crusade burned that down as well.

In the following centuries, the Spanish would burn the Mayan Codices, the British would burn the American Library of Congress, and Nazis would burn libraries across Czechoslovakia, Yugoslavia, Poland, France, and even Germany (see Figure 37-1). Even today, the shrill and familiar sound of smoke alarms may be heard ringing throughout conflict zones. History may be written by the victors, but its transcripts are flammable.

Figure 37-1. Plaque in Berlin reading, "In the middle of this square on 10 May 1933, national socialist students burned the works of hundreds of writers, journalists, philosophers, and scientists."[4]

[3]Cartwright, Mark. "1204: The Sack of Constantinople." Ancient History Encyclopedia. June 08, 2018. Accessed June 09, 2018. https://www.ancient.eu/article/1188/1204-the-sack-of-constantinople/.

[4]Blok. Berlin Plaque Book Burning. Digital image. Pixabay. June 25, 2014. Accessed June 7, 2018. https://pixabay.com/en/berlin-plaque-book-burning-376449/.

Modern-day digital repositories fare no better. Hurricanes, viruses, Trojans, malware, hacking, hardware failures, and simple human error destroy millions of records each year.

When New Orleans' levee system failed in 2005, floodwaters covered approximately 80% of the city.[5] Hurricane Katrina killed 986 residents and caused irreparable harm to the city's infrastructure, including the widespread loss of data at Louisiana's vital records office. A century's worth of birth certificates, marriage licenses, and other forms of identification were swept away, leaving thousands of people with neither a history nor an identity.

Malicious hacking rivals any natural disaster. Some attacks merely intercept data, while others deliberately destroy it. In 2014, the code-hosting company Code Spaces suffered a massive data loss[6] after a hacker deleted the company's cloud storage, onsite disks, and offsite backups. This attack—a 12-hour long digital siege—destroyed the company, forcing Code Space to close its doors that same year, proving that no data is immune from deletion.

Along with deletion comes a new threat—encryption. Encryption-based ransomware, such as CryptoLocker,[7] threatens institutions and individuals alike, from hospitals to hairstylists. Although cryptography does not destroy data, it may lock data away behind a nearly impenetrable cryptographic algorithm. Without payment of a ransom, the data disassembles into irretrievable bits; the digital equivalent of a book burning.

Books burn, documents vanish, systems are hacked. We should be amazed that any recorded history exists at all.

History has but one protection: when we document, we breathe life back into history. We recount stories of queens, poets, CEOs, and fry cooks. We narrate how people work and play. We tell new tales, creating a new historical record, completing one more loop in an infinite cycle that reveals and preserves the human experience.

Documentation begins with a name. So, let us start there.

---

[5] Plyer, Allison. "Facts for Features: Katrina Impact." The Data Center. August 26, 2016. Accessed June 9, 2018. https://www.datacenterresearch.org/data-resources/katrina/facts-for-impact/.
[6] Venezia, Paul. "Murder in the Amazon Cloud." *InfoWorld*. June 23, 2014. Accessed June 09, 2018. https://www.infoworld.com/article/2608076/data-center/murder-in-the-amazon-cloud.html.
[7] "CryptoLocker." Wikipedia. June 08, 2018. Accessed June 09, 2018. https://en.wikipedia.org/wiki/CryptoLocker.

Naming

In 1826, John Walker changed the world with the flick of his wrist. He invented the deceptively simple, yet incredibly practical "friction light."[8] Its design made starting a fire nearly effortless, for a mere swipe across a rough surface would produce a flame. In comparison, the going alternative at the time—striking flint and steel to produce a spark—required both skill and patience; two commodities rarely available to the busy cooks, freezing pioneers, and late-night bookworms of the day. Walker's invention would go on to ignite Olympic torches and enflame political revolutions. Yet despite it playing an integral role in many moments in history, the friction light faded into obscurity. Perhaps John Walker's legacy would have endured if he had given his product a better name.

Countless innovators and marketers improved on Walker's original design. A mere three years after its invention, Samuel Jones created his "Lucifers," which arguably had the most fitting product name ever devised. Its main ingredient, white phosphorus, was prone to explode in the hands of its users as well as rot the bones of its manufacturers. Subsequent companies renamed the product, adding and tweaking features, improving safety and usage along the way.

Today, we call this device a match. You can find decorative matches, waterproof matches, stormproof matches, safety matches, long-reach matches, scented matches, and an entire subculture of match collectors known as phillumenists. Walker's single innovation evolved into thousands of offshoots made by hundreds of companies. We see the same with many inventions, from dumbbells to smartphones. One core idea leads to innumerable variations.

The Need for a Unique Name

We identify much of the modern world through unique names. We access uniquely named websites, such as Amazon.com. We send emails to uniquely named addresses, like book@edwardstull.com. We use uniquely named operating systems, like Android Lollipop. These names encapsulate their differentiation. Website names represent strings of numbers using the Internet Protocol; email address names represent mailboxes as defined by RFC 5321 and 5322;[9] operating system names represent the versions of software that sit between devices and applications. A name's quality lies in its uniqueness: a single name that is never repeated.

---

[8] Walker, John, and Doreen Thomas. *The Day-book of John Walker Inventor of Friction Matches: Annotated Extracts.* Cleveland, Middlesborough: D. Thomas, 1981.
[9] "Email Address." Wikipedia. June 06, 2018. Accessed June 09, 2018. https://en.wikipedia.org/wiki/Email_address.

A search for "matches" on Amazon returns over 300,000 products, covering everything from stormproof matches to handmade candleholders. Amazon must keep track of every single one. To solve this complex tracking problem, Amazon gives each product its own identifier, the ASIN. Amazon's Standard Identification Number[10] is a unique 10-digit string of letters and numbers. Amazon uses a 10-digit string to manage 480 million products, including every style of matches.

Naming computer files presents some of the same challenges. Same-named, nearly named, and vaguely named files mix like flames in a bonfire. Naming a file "final" only indicates the date at which you thought you were finished. But finality often changes over time. Today's "final" becomes tomorrow's previous draft.

Names fail when we rely on other forms of identification, such as creation and edit dates. Emailing and uploading may strip embedded file date information, rendering a file's date to the day it was last sent, received, or downloaded.

You may think a file's directory structure connotes its meaning (e.g., Acme/2017/ presentation.doc). Deriving a file name based on its temporary location is fraught with problems. Once moved outside of its intended directory, the file's meaning becomes untethered. Your "presentation.doc" could represent a presentation for any person, place, or thing, at any time.

What Makes a Good Name?

Several years ago, I met Neil Kulas while working at a small agency in Milwaukee. His smart approach to naming conventions has kept me organized ever since.

Neil's naming convention is as follows:

YYYY-MM-DD-HHMMa_company_project-name.ext

For example:

2017-04-15-230pm_zippo_presentation.doc

With such a name, you can upload, download, and email to your heart's content. Starting with the date and time, you ensure the uniqueness of the file name, allowing you to quickly see which file is the most recent. The name will always preserve its date. For example:

2017-04-15-230pm_zippo_presentation.doc

2017-04-15-231pm_zippo_presentation.doc

2017-04-14-232pm_zippo_presentation.doc

---

[10] "What Are UPCs, EANs, ISBNs. And ASINs?" Amazon. Accessed June 09, 2018. https://www.amazon.com/gp/seller/asin-upc-isbn-info.html.

Company names can be the quickest way to differentiate files if you have multiple clients, partners, vendors, or competitors. For example:

2017-04-15-230pm_**zippo**_presentation.doc

2017-04-15-230pm_**bic**_presentation.doc

A short description eliminates the guesswork when determining a file's contents. For example:

2017-04-15-230pm_zippo_**presentation**.doc

2017-04-15-150am_zippo_**agenda**.doc

2017-04-14-530pm_zippo_**contract**.doc

A name may be the first experience a person encounters, be it the name of a product, proposal, or PDF. An arbitrary name steals viewers' valuable time as they struggle to understand its meaning. They gaze and wonder: what is this, when was it created, who is it for? A good name answers these burning questions, thereby creating clarity for viewers and illuminating the path ahead.

Fidelity

In Plato's allegory of the cave, a group of prisoners sits chained within a deep cavern. Unknown captors guard over them with oppressive precision, ruling every aspect of the prisoners' lives, determining what they see, hear, and do. It is a plight every user endures.

The prisoners know no other life, for they were born into captivity. They sit day and night, staring at the cave's back wall. Behind them roars a mighty fire. However, the prisoners never see it because tightly bound chains prevent them from turning around. They view the fire's illumination and its flickering shadows. The prisoners gasp in awe when, in front of them, the silhouettes of beasts and men appear to emerge out of the darkness and dance across the once blank cave wall. All the while, the prisoners do not realize that the silhouettes are shadow puppets controlled by their captors. To the captors, the trickery is merely a cruel form of amusement. To the prisoners, the shadows are as real as any other experience.

Figure 37-2. Plato's Allegory of the cave[11]

To the average person, user experience appears much like the dancing shadows in Plato's allegory. People see the shadows of UX cast upon software interfaces: buttons, links, labels, and the like. However, many people do not understand from whence UX originates—they never see the proverbial fire.

From applied psychology to user testing, several subjects illuminate user experience: human factors, design research, information architecture, human–computer interaction, cultural anthropology, service design, and usability engineering. All play a part. Without at least a glimpse into such knowledge, people may misinterpret how software is designed. They see an interface and mistake it for an experience.

Like prisoners watching the silhouettes cast upon a wall, people recognize the familiar. They look up and say, "I know what that is: that's a button, that's a screen, and that's a website!" Yet, none of these observations indicates a user's experience. A button shows shape and color, but it does nothing to demonstrate a user's behavior. A screen presents graphic design and copywriting, but it does nothing to reveal a user's perception. A website displays content and navigation, but it does nothing to signify a user's context or where she will travel next.

---

[11]Jan Saenredam, "Plato's Allegory of the Cave," 1604, public domain.

When we document UX, we describe an experience. Flowcharts indicate the interconnections between people, places, and things. Wireframes diagram functional displays, such as dialogs and screens, but these, too, are mere representations of a future reality. Understanding these documents becomes a matter of fidelity.

Fidelity is the degree of sameness between two things. A high-fidelity document mirrors a future product, whereas a low-fidelity one merely references it. We often prefer to see high-fidelity documents because they demand less of our imaginations—we see what *will be*, rather than what *could be*. But we must ask ourselves: when is high fidelity necessary and when is it distracting?

Necessary and Distracting Fidelity

High fidelity at all times is both an impossible and unwise pursuit. It requires an unattainable level of effort, as every idea needs to be designed with precision, built to completion, and tested to perfection. To understand an experience, we need only reach a level of fidelity that communicates what is necessary at any given moment.

Consider the following example:

> A large, red button with 14px yellow text reads "Complete Purchase" in Helvetica Neue Bold.

What part of this statement is necessary? Well, it depends. If the button were mocked up days before a website launch, perhaps everything about the statement would be necessary. Size, color, and font may be required to implement it. But, if the button were hand-drawn on a cocktail napkin during a project's first week, perhaps much of this statement would be distracting. A live website is high fidelity; a cocktail napkin is low fidelity. Yet, both communicate what is necessary at their respective points in time.

Choosing the right fidelity for the right purpose at the right time is an art in itself. We have many choices. However, almost all UX documentation falls into one of three general categories: maps, mock-ups, and prototypes.

Maps

A map defines the hierarchies and interconnections within a digital experience. It shows the respective connections between areas (e.g., screens), indicating the movement of users or processes. Flowcharts, site maps, journey maps, and knowledge maps all fall under this category. We will discuss several of these in subsequent chapters.

Mock-Ups

Mock-ups create a visual reference of elements appearing within a digital experience. A low-fidelity mock-up is a wireframe (see Figure 37-3). Often, we wireframe elements that appear on a screen (e.g., an app's sign-in). Wireframes indicate the approximate placements of content and functionality, such as buttons and dialogs. However, wireframes do not indicate color, typography, photography, illustrations, or copywriting. A high-fidelity mock-up—a design mock-up—picks up where a wireframe leaves off, including items previously not indicated. Design mock-ups frequently mirror the final appearance of a future digital experience.

Figure 37-3. A sketch compared to its respective low-fidelity wireframe

Prototypes

Prototypes emulate a future digital experience. Paper-based prototypes serve as quick, low-cost emulations—somewhat like a puppet show. They require the movement of physical pages of paper to mimic the functions of a working application or website. You can create a paper prototype with a printed set of wireframes or design mock-ups. Digital prototypes range from linked images to coded interfaces. A high-fidelity digital prototype may be nearly indistinguishable from a fully functioning product, rivaling it in both complexity and build effort.

The choices among high- and low-fidelity approaches can lead UX teams to over-design. Some UX teams present high-fidelity mockups of every screen, button, and link. They believe project stakeholders cannot appreciate low-fidelity documents. However, like Plato's prisoners, stakeholders may misinterpret UX for the shadows it casts. In the pursuit of high fidelity, they forget the underlying experience—they forget the fire.

Showing Fire

We are suckers for appearances. Robert Cialdini wrote in his 1984 book *Influence: The Psychology of Persuasion*, that the more attractive something looks, the more successful, smart, and trustworthy it may appear. A fancy suit connotes credibility as much as a diploma. Appearance engrains itself into our daily lives, from the clothes we wear, to the cars we drive, to the products we buy. It is expected—if not entirely natural—to evaluate documentation in the same way.

Take a handful of people off the street and ask them to view a low-fidelity UX document, such as a wireframe or flowchart. You will hear responses such as "it looks unfinished," "it looks technical," and inevitably, "it looks ugly." While these reactions are predictable, they stem from the way in which the question was asked. A low-fidelity wireframe will always lose to a high-fidelity mock-up when viewed for its aesthetic merits. To demonstrate the merits of low fidelity, we must remove distractions and let the experience shine through.

For instance, let us revisit our prior example (see Figure 37-4):

Complete Purchase

Figure 37-4. Complete purchase button

> A large, red button with 14px yellow text reads "Complete Purchase" in Helvetica Neue Bold.

> ...could instead be described as the following (see Figure 37-5):

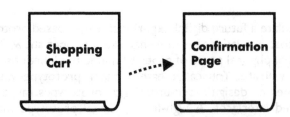

Figure 37-5. Shopping cart to confirmation page flow

> Upon clicking the "Complete Purchase" button, the user views the confirmation page.

The information we communicate here is sparse. We do not include the size, color, or font. Instead, we replace the description of what we see with what the user does. We highlight the user's experience rather than our own.

Effective UX documentation directs our gaze toward the user's experience, transforming the superficial into the meaningful, freeing us from our delusions and illuminating the shadows that dance across our screens.

Key Takeaways

- Millions of records are destroyed each year by natural and human-made causes.

- We identify much of the modern world through unique names.

- A name may be the first experience a person encounters, be it the name of a product, proposal, or PDF.

- Today's "final" becomes tomorrow's previous draft.

- The appropriate level of fidelity is whatever communicates the necessary at any given moment.

- High fidelity at all times is both an impossible and unwise pursuit.

- Interfaces are not experiences.

Questions to Ask Yourself

- What UX documentation is required for each team member to do her or his job?

- Do remote team members require additional UX documentation to compensate for communication challenges (e.g., disparate time zones, nonnative languages, or varied national holidays)?

- Would the file name of a file be descriptive enough to locate it again a year from now?

- Does a document's fidelity help or hinder its use?

Personas

With its feathered haircuts, tight-fitting pant suits, and abundant chest hair, the long-running game show, *The Dating Game*, filled American TV screens and living rooms with bawdy singles and hopeless romantics for over three decades. It ran from 1965 to 2000. It started in the Age of Aquarius and ended in the Internet Age, thereby amounting to the longest-running, most-viewed study on dating habits, people, and personas.

The show's format was simple: an attractive date-seeker sat behind a wall and asked potential bachelors their opinions on dating. Thought-provoking questions such as, "What would you do to show this girl a good time?" were met with equally cringe-worthy answers, ranging from promises of foot rubs to promises of stalking. Through a cloud of lewd exchanges and the haze of Jovan Musk Oil, a winning contestant would finally emerge. Studio audiences would laugh and revel at the date-seeker's apparent surprise in her choice of bachelor. Frequently, eager faces showing anticipation were drained of excitement and immediately replaced by apprehension upon the reveal.

We could all empathize with the plight of both the date-seeker and contestant. After all, imagining an ideal mate is a difficult exercise. Give it a try. Is she or he older than you? How does this person spend her or his day? Perhaps she's a chemist. Perhaps he's a carny. Is she serious? Is he funny?

Your responses may describe a remarkable person: someone who recites poetry while chopping down trees; someone who looks like a runway model but works as an astrophysicist. You may describe an ordinary person: someone who listens to sports radio while mowing the lawn; someone who looks like an accountant and works as… an accountant. Regardless of whom you describe, you describe the attributes and behaviors of a human being. You describe a persona.

© Edward Stull 2018
E. Stull, *UX Fundamentals for Non-UX Professionals*,
https://doi.org/10.1007/978-1-4842-3811-0_38

An ideal persona is based on a real person whose attributes and behaviors match a particular population's. However, an individual person rarely reflects the diversity of an entire population, leading us to make sampling errors (see Chapter 33, "Quantitative Research"). Lumberjack poets and astrophysicist-supermodels be damned.

Often, a demographic persona more faithfully represents the entirety of a population. A common criticism of such personas is that they are make-believe. Although the people we describe may be figments of our imagination, everything about them is based on facts. Similar to our ideal mate, a persona may only exist within the confines of our descriptions. However, we can still imagine how this person would interact within our physical world. Our thespian lumberjack may be a man who enjoys beat poetry by day. Our Heidi-Klum–Stephen-Hawking lovechild may be a woman who counts the stars at night. We prove or disprove these attributes and behaviors through research data.

A demographic persona is a human-shaped container of data. In this container, we place the attributes of a person, such as age, gender, and annual income. We also place behaviors, such as "recites poetry," "chops down trees," and "shops for tight-fitting pantsuits." How do we determine which attributes and behaviors to include? We only include data supported by research. Anything short of research is merely a daydream.

For example, according to the Bureau of Labor Statistics, lumberjacks tend to be male and around 45 years old (97.9% male workforce, median age 44.9[1]). Furthermore, the *2003 National Assessment on Adult Literacy* shows a high literacy rate among 40–49 year olds.[2] So, we now have support for the likelihood that our thespian lumberjack is male and can read. Yet, when we compare these numbers to the National Endowment for the Arts' *2012 Survey of Public Participation in the Arts*, we see that only 6.7%[3] of Americans have read a book of poetry in the past year. Finding a lumberjack that recites poetry while chopping down trees would indeed be remarkable—perhaps even statistically improbable. Research on our astrophysicist-supermodel may lead us to strong evidence of a correlation between astrophysics and fashion modeling, or disprove it entirely. Either result requires research data; our biases, especially as they pertain to gender and social class, are too pervasive to go unchecked.

---

[1] U.S. Bureau of Labor Statistics. Accessed May 28, 2018. http://data.bls.gov/cgi-bin/surveymost.

[2] "National Assessment of Adult Literacy (NAAL)." Revenues and Expenditures for Public Elementary and Secondary Education: School Year 2001-2002, E.D. Tab. Accessed June 09, 2018. https://nces.ed.gov/naal/health_results.asp#AgeHealthLiteracy.

[3] "National Endowment for the Arts Presents Highlights from the 2012 Survey of Public Participation in the Arts." NEA. January 12, 2017. Accessed June 09, 2018. https://www.arts.gov/news/2013/national-endowment-arts-presents-highlights-2012-survey-public-participation-arts.

Remarkable personas are often referred to as an "edge-case": a persona that may be true but is often statistically insignificant. When we develop personas, we wish to describe the significant attributes and behaviors of a population, but we must first decide if the persona will be aspirational or historical.

Historical versus Aspirational Personas

An often-overlooked part of creating personas is determining if a persona is historical or aspirational. Historical personas comprise users who used a product in the past. Aspirational personas comprise potential new users. This differentiation may seem to be trivial, but it forms the basis for many future decisions.

For example, let's say you created a dating app. Your app allows users to quickly find eligible singles within a 500-foot radius. The app is a few years old, and you have a wealth of analytic information about its use. Research data may indicate the following attributes and behaviors of a prior user:

- 18–34-year-old female

- Lives in high-population urban area

- Frequently visits nightclubs and dines at restaurants

This historic persona may align well with the app you created. This persona indicates a young woman who frequently finds herself within a 500-foot radius of other eligible singles, as she travels between nightclubs and restaurants in an urban setting.

Now, let's create an aspirational persona. We wish to reach a user with the following attributes and behaviors:

- 18–34-year-old male

- Lives in medium-population suburban area

- Occasionally dines at restaurants

Our research data may show that men matching these attributes and behaviors exist; therefore, our aspirational persona is valid. However, this hypothetical person would likely interact with our dating app in very different ways than our historical persona. The aspirational persona points to a young man who rarely finds himself within a 500-foot radius of other eligible singles because he dines at restaurants in a suburban setting only occasionally. He may never realize the complete benefits of your app.

The example used three attributes to indicate a potential pitfall: a persona may never realize the app's benefits. But, we rarely get a full picture of a person through such limited information. Just like a real person, a persona becomes more alive to us the more we know about it. A persona describes a life.

So far, our personas only include a few attributes and behaviors. Age and location are included in many personas, but these are only the start (see Figure 38-1). Access to healthcare, business ownership, citizenship status, country of origin, device ownership, disability, education, employment status, ethnicity/race, sex, gender identity, gun ownership, home ownership, income, languages spoken, languages spoken in the household, marital status, military service, number of children, number of siblings, pet ownership, political party affiliation, religious denomination, religious service attendance, sexual preference, shopping habits, social networking usage, technical aptitudes, trust in government, union membership, and vehicle ownership, to name just a few.[4] If episodes of *The Dating Game* had covered such an exhaustive list of data, they wouldn't have ended with so many surprises.

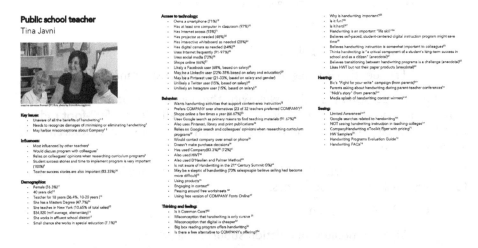

Figure 38-1. An example persona containing a full range of demographic and behavioral data[5]

[4]Bureau, US Census. "Data." Historical State Tables. November 01, 2015. Accessed October 12, 2015. https://www.census.gov/data/tables/time-series/demo/families/states.html.

[5]Ilmicrofono Oggiono. "Students-in-class-with-teacher-reading." Digital image. Flickr. January 29, 2014. Accessed June 7, 2018. (CC BY 2.0) https://www.flickr.com/photos/115089924@N02/12212474014/.

Exaggeration and Accuracy

If you've ever joined an online dating service, you've experienced the process of creating a persona. It was your dating bio. In it, you likely described a few of your attributes: your age, your height, and an optimistic approximation of your weight. You ran a spell check. You uploaded the best picture of yourself. You probably also described several of your behaviors: what you do in the morning, during the day, and at night.

Now imagine the bio you would write if you weren't limited by reality. You recite poetry in the morning. You work as a runway model during the day. You map the stars at night. You are your ideal age. You are the perfect height and weight. You are exactly as you wish to be. It is a fun exercise to do, but perhaps not one backed by evidence.

You may enjoy poetry but only read it a few times a year. Your modeling career is limited to posing in the bathroom mirror. You can peer into the night sky and find the Big Dipper—or was that the little one? Reality has a way of making us all remarkably average. This is our real bio; this is our persona.

Because software creators may use personas to support their work, attributes and behaviors are sometimes inadvertently—or intentionally—exaggerated to make personas appear more attractive. Incomes are increased. Ages are adjusted. Engagement is elevated. Interest is enhanced. Once historical personas inflate into aspirational ones. After all, who wouldn't want to be a runway model that moonlights as an astrophysicist? However, we should avoid such exaggerations, not only because such personas are often inaccurate, but also because this inaccuracy bleeds into the resulting design of the software products we build.

A common exaggeration is people's interest in a new product. It often manifests itself in the extraordinary time and effort they will commit to a product that offers them no immediate value. For example, the belief that people inherently want to explore websites. They don't. The default position toward most websites in a person's mind is apathy. A person is often no more likely to explore a website than they are to explore a dark room. You need to shed a bit of light on something or else people simply close the door and walk away. Personas should reflect apathy as much as they reflect interest, because only then can we accurately estimate how to change these default positions into more beneficial behaviors. If we exaggerate, we will never design solutions that accurately meet users' needs. We need to know where people stand before we guide them somewhere else.

When we design experiences, we must always consider the personas of users. They experience the world, as well as what we create for them, in different ways. For an experience may delight one person, outrage a second, and go unnoticed by a third. We sometimes find ourselves sitting behind a wall of ignorance, blinded by the stage lights of our projects' demands, unable to see users for whom they are and what they truly want. The questions we ask—and the answers personas give—allow us to imagine these hypothetical people and fulfill their needs. Will they fall in love with what we create, or will the romance never start? The answer often lies in our ability to transform research into understanding, and a persona into a person.

Key Takeaways

- A demographic persona is a human-shaped container of data.

- Personas not backed by evidence are daydreams.

- Decide if a persona is historical or aspirational.

- Avoid exaggeration, especially as it pertains to interest in your product.

Questions to Ask Yourself

- Is the persona historical or aspirational?

- Have I exaggerated positive attributes of the persona?

- Have I deemphasized common and negative attributes of the persona?

- Is the persona representative of the target population?

- Can I find real people within the target population who match the persona?

- Does the persona(s) account for the diversity contained within a target population?

Journey Mapping

Far off in the Pacific Ocean, past thousands of miles of rolling waves, beyond the western coast of the Philippines, sit the Spratly Islands.[1] Less than two square miles of visible land stipple 200,000 square miles of open water. Like paint droplets sprayed across the canvas of the South China Sea, dozens of uncharted reefs and uninhabited atolls form what sailors call the Dangerous Ground.

It is an ominous label for what amounts to some rocks, sand, and few palm trees. Despite six nations asserting ownership over the islands, no single nation, no single source, knows the exact boundaries or composition of the area. Centuries of cartography have yet to precisely map the Spratly Islands. Without an accurate assessment, navigational charts disagree, militaries skirmish, and ships run aground. The islands remain a mystery.

As with any exercise in discovery, be it an island chain or user experience, a map helps guide our way. Maps transform abstract notions into tangible landmarks, clarifying foggy visions, becoming the solid earth to which we anchor our ideas. We uncover dangers hiding beneath the surface of an experience—the obstacles that compromise business objectives and force users to abandon. We avoid territorial disputes between team members, designing software based on a user's journey rather than our own.

---

[1]Hancox, David, Clive H. Schofield, and John Robert Victor. Prescott. *A Geographical Description of the Spratly Islands and an Account of Hydrographic Surveys amongst Those Islands.* Durham: International Boundaries Research Unit - Univ. of Durham, 1995.

© Edward Stull 2018
E. Stull, *UX Fundamentals for Non-UX Professionals*,
https://doi.org/10.1007/978-1-4842-3811-0_39

When we map user experience, we do not describe a fixed, unchanging landscape. Its borders are far more dynamic. It expands and contracts based on the emergence of technologies, the volatility of markets, and the whims of human beings. To chart the course ahead, we design an experience before it begins.

Journey Mapping

Even with the best research, we only partially understand our audiences. Eventually, we learn their ages and incomes, likes and dislikes, proclivities and peccadillos. Although this information shapes our creations, it does not tell us what to create. To create, we must design an experience.

How did users get here? What are they doing now? Where will they go next? We ask ourselves these questions when we create—be it building a website or writing a book.

You, the reader, are a bit of a mystery to me. As a writer, I am left wondering how you arrived at this exact moment. Perhaps you turned a printed page, followed a blog link, or fast-forwarded an audio narration. One of those activities led you here, but we both know your journey started well before that.

We can map this journey, starting from your birth and ending with this moment (see Figure 39-1).

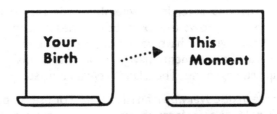

Figure 39-1. A simple journey

This journey is straightforward as it follows a linear path from A to B. However, users have three states of existence: where the user was, where the user is, and where the user is going. We can add these states to our diagram. For demonstration purposes, let us assume you will read the next page of this book (see Figure 39-2). By doing so, we address where you are going.

Figure 39-2. Where the user is going

This journey now shows the before, during, and after of your current experience. With this three-step framework, we can map out every conceivable user experience. For example, we can map the journey a person may have taken if he or she read a blog post (see Figure 39-3).

Figure 39-3. Before, during, and after

Now, imagine you want to map all other possible journeys that someone may take to reach this moment. Such a map becomes considerably more complex.

- Read a printed page
- Read a blog post
- Read an email
- Read a quote containing this text
- Read a tweet
- Read a Facebook post
- Read an Amazon review
- Read the text signed in American Sign Language
- Read using an assistive technology
- Listen to an audio narration

How to Create a Journey Map

A whiteboard or wall serves as an ideal location to engage in journey mapping exercises—the bigger, the better. Cover its surface with a roll of craft paper or sheets from an easel pad. (This allows you to roll up the journey map once it is finished.)

Journey mapping exercises do not require a specific number of meeting attendees. Five people works well, but you could also create a journey map alone or with a large crowd. Include attendees who have specialized knowledge about a business or audience, such as board members, call center staff, lawyers, and database administrators.

Your journey map will detail a linear timeline. Many people use a marketing lifecycle, but you can use any sequence of events: a path to purchase, a cruise ship check-in process, stages of writing a book, etc. (see Figure 39-4).

Consumer Journeys

Personae	Scenarios	Motivators	Engagement Channels	Conversion Points	Retention
• Consumers	• Researching who COMPANY is	• Authority	• COMPANY .com, Company Info	• Contact form	• Familiarity
• Media	• Looking for CPG brand Info	• Efficacy	• COMPANY .com, Our Brands	• Social share	• Customer service
• Investor/Influencer	• Seeks balanced information	• Convenience	• COMPANY .com, Our Commitments	• Commenting	• Brand loyalty
• Employees	• Seeks response to viral negatives	• Food Safety	• COMPANY Stories	• Phone call	• Quality of content
	• Researching company policy/stance	• Environmental	• Facebook	• File download	• Fresh content
	• Seeks environmental policies	• Water use	• Twitter	• Email-to-friend	
	• Wishes to complain	• Water contamination	• Pinterest	• Attitude/sentiment change	
	• Looking for article ideas	• Energy	• LinkedIn		
	• Looking for open positions	• Workers issues	• Youtube		
	• If COMPANY was in news...	• International concerns	• Instagram		
		• Ratings/Reviews	• Email		
		• Competitive interest	• Blogs		
			• Reddit AMA		
			• Google Hangouts		
			• Vimeo		
			• Vine		
			• Public speaking / campus outreach		
			• Online QA and archives		
			• Phone		
			• Contact form		

Figure 39-4. An example journey that maps a customer's initial motivation to sustained retention

For demonstration purposes, let us use the typical marketing funnel of an e-commerce website, starting with awareness and ending with retention. Divide the wall into several sections: Awareness, Acquisition, Conversion, and Retention. To start, we ask attendees an open-ended question:

What are our business objectives?

The attendees might answer:

"The company wants users to buy products."

"The company wants users' email addresses."

We write each business objective on a Post-It note and place it under the "Conversion" heading (see Figure 39-5).

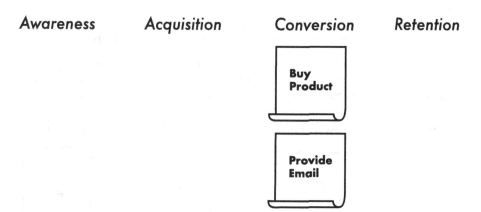

Figure 39-5. Business objectives under conversion

Getting a person to do anything is a conversion. Although the term often describes an exchange of money, it is actually about the exchange of value. We want something from a user, so we need to give something back. (Hold that thought. We will revisit conversion in a minute.)

What do users need?

"Users want to research a product."

"Users want to buy a product."

"Users need to review their past orders."

Research is a form of acquisition. So, we place "Research Product" under "Acquisition" (see Figure 39-6).

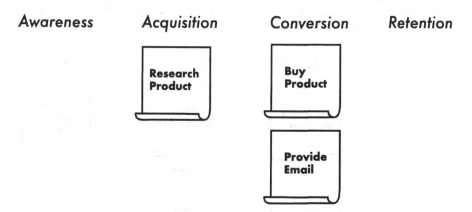

Figure 39-6. User goal under acquisition

Buying a product is a form of conversion. We have already placed "Buy Product" within the journey because it is also a business objective.

Reviewing a past order falls under retention. Retention describes our ability to maintain an ongoing relationship with users. For example, we retain users who return to our website to review their past purchases (see Figure 39-7).

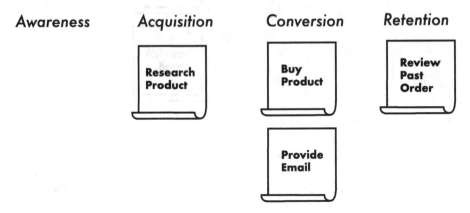

Figure 39-7. User goal under retention

At this point in the exercise, some attendees may start questioning how all the Post-Its correlate. This is where journey mapping proves its worth. We start connecting points on the map.

A minute ago, we discussed conversion being an exchange of value. The business wants to sell products; a user wants to buy products. This exchange is clear—so, we are good there. However, the business also wants users' email addresses. What does a user get in return? We need to offer them something of value. Perhaps in exchange for an email address, we offer a 10% discount on a user's first purchase (see Figure 39-8).

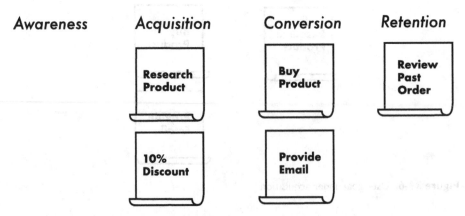

Figure 39-8. Discount leads to conversion

Before any journey begins, users need to become aware of it. Perhaps users learn about a product through a Google search. Perhaps users are reminded of their past purchases through a targeted email. Write each one down on a Post-It and place them under "Awareness" (see Figure 39-9).

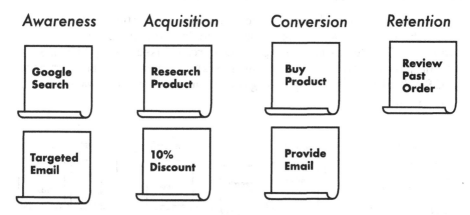

Figure 39-9. Awareness leads to acquisition

Of the four categories, retention is most often overlooked. Retention may be the starting point of a previous user's journey. For example, a business may wish to offer a rewards program or exclusive discounts to repeat customers. Even something as small as a thank-you message can help retain users. Place each tactic under "Retention" (see Figure 39-10).

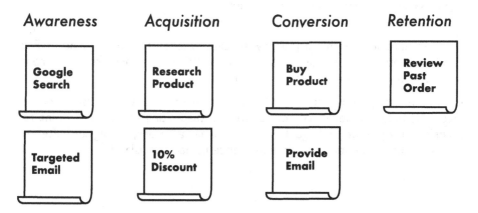

Figure 39-10. Retention may be a starting point

The secret behind journey mapping is ensuring that all points connect: each awareness point connects to an acquisition, each acquisition point connects to a conversion, each conversion point connects to a retention. Like a bridge joining two islands, each pairing allows users to move from one part of an experience to the next.

In our final step, we look for ways to strengthen connections among points within the journey map. For example, we could add a custom landing page between a Google search and a product purchase (see Figure 39-11). Such a tactic links a search term (e.g., sunglasses) to a respective product page (e.g., Ray-Ban), thereby strengthening the connection between awareness and conversion.

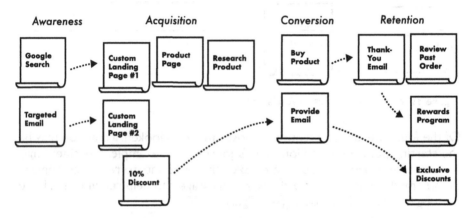

Figure 39-11. A complete, interconnected journey map

Weak connections within a journey indicate where users may abandon. Pay special attention to these areas. If you cannot think of a reason to move from one point to next, neither can your users.

Journey mapping shows us that UX is far more than a set of disconnected tactics. It entails a series of connections, transporting users from point to point as they pursue their goals. Each point has a before, during, and after—nothing happens on its own. No experience remains an island.

Key Takeaways

- Even with the best research, we only partially understand our audiences.

- We can map out every conceivable user experience by understanding the before, during, and after of an experience.

- Weak connections within a journey indicate where users may abandon.

- If you cannot think of a reason to move from one point to next, neither can your users.

Questions to Ask Yourself

- Who is having the experience?

- What steps do users take to complete their goals?

- How can I support users at each step within their journey?

- Am I providing too much help to users?

- Where do user goals intersect with business objectives?

- What happens before, during, and after each step within an experience?

- Where are the holes within an experience?

- Where is additional research needed?

- How do I retain users once they have completed their goals?

Knowledge Mapping

On a typical school night in the 1980s, you might have found a kid slumped over a volume of the *Encyclopedia Britannica*. The 30-odd leather-clad books inspired countless children's science reports and history essays. Each volume presented a slice of the world's collective knowledge, alphabetically labeled from Accounting to Zoroastrianism.

Coincidentally, in 1984, United Artists released the hit movie, *Red Dawn*, in which Russia and Cuba invade the United States. The film was staggeringly jingoistic, even when seen through the polarized lens of the Cold War. My 13-year-old self fantasized about how I would defend the American way of life from a communist invasion, principally by burying the encyclopedias in the backyard for safekeeping. After all, who would remember how to do important things like build a dam after World War III? I was certain that river waters would stay at bay as long as I had page 440 of *Macropaedia, Knowledge in Depth 5 (Conifer—Ear Diseases)*. Each volume was segmented according to topic, and each topic was divided into endless subtopics. You could spend an entire post-apocalyptic lifetime perusing its pages. In retrospect, why I thought the Russians would want a suburban Ohio kid's encyclopedias—written in English, no less—is lost on me, but that did not keep me from daydreaming about protecting my *Britannicas*.

© Edward Stull 2018
E. Stull, *UX Fundamentals for Non-UX Professionals*,
https://doi.org/10.1007/978-1-4842-3811-0_40

Even at a young age, we learn that information is valuable. We also learn that acquiring it often proves be a challenge. Knowledge takes time and effort. As we grow older and our professional lives become more demanding, we sometimes reach for a quick solution rather than a well-thought-out one. Like sleepy-eyed schoolchildren trying to finish their homework, we sacrifice knowledge in order to save time.

Knowledge Mapping

Speed and knowledge are mortal enemies; yet, this standoff preserves a balance in our professional world. Slow-moving companies risk being overtaken by their faster challengers; unknowledgeable companies risk being overtaken by their smarter competitors. We reach a detente when we quickly map the boundaries of our understanding.

Knowledge mapping is a deceptively simple technique. We tackle large goals by breaking them down into smaller parts.

Let us imagine we work at a library. Our goal is to "assist library visitors." To create a knowledge map, we start by clarifying the nouns within our goal.

Visitor:

- **Member**

- **Non-member**

As you might expect, we can break each topic down further.

Visitor:

- Member

- - **Active**

- - **Expired**

- - **Provisional**

- Non-member

- - **Prospective**

- - **Guests**

- Staff member

- - **Librarian**

- - **Customer service**

- - **Public relations**

- - **Accounting**

- - **Network administration**

- - **Security**

Now, we look for any verbs within a goal and ask yourself what each means. The verb "assist" in the context of a library might mean the following:

Assist library visitors:

- *Locate a book*

- *Locate a DVD*

- *Find a place to eat nearby*

- *Validate parking*

We now have a list of topics to break down into subtopics. Again, describing each verb uncovers additional meaning:

Assist library visitors:

- Locate a book

  - - *Find by book title*

  - - *Find by author name*

  - - *Find by ISBN*

- Locate a DVD

  - - *Find by DVD title*

  - - *Find by release date*

  - - *Find by genre*

- Find a place to eat nearby

  - - *Display list of restaurants*

  - - *Display cafeteria hours*

  - - *Display vending machine locations*

- Validate parking

  - - *Stamp parking ticket*

With another pass, we can further refine each sub-subtopic:

Assist library visitors:

    - Locate a book

        - - Find by book title

        - - Find by author name

        - - Find by ISBN

            - - - Refer ISBN inquiries to customer service desk

    - Locate a DVD

        - - Find by DVD title

        - - Find by release date

        - - Find by genre

    - Find a place to eat nearby

        - - Display list of restaurants

            - - - Display location on map

            - - - Display phone numbers of local cab companies

        - - Display cafeteria hours

            - - - Display walking directions to cafeteria

            - - - - Request wheelchair assistance

        - - Display vending machine locations

            - - - Display walking directions to vending machines

            - - - - Request wheelchair assistance

    - Validate parking

        - - Stamp parking ticket

Although this example may look like a simple outline, it serves to describe knowledge, not content. Knowledge maps discover correlations that were not previously apparent. Consider the following:

"Assist library visitors" means...

1. *Find a book by title, author name, and ISBN*

2. *Find a DVD by title, release date, and genre*

3. *Display a list of restaurants and their locations*

4. *Display vending machine locations*

5. *Display phone numbers of local cab companies*

6. *Request wheelchair assistance to cafeteria and vending machines*

7. *Display cafeteria hours*

8. *Stamp parking ticket*

We transform the large goal of "Assist library visitors" into specific user tasks and stories.

Knowledge maps preserve the balance between speed and knowledge by recasting a single large goal into many smaller ones. Small goals form the basis of modern development methodologies, such as Agile and Lean. Goals become more specific, practicable, and achievable. A little knowledge can be a wonderful thing.

Key Takeaways

- Knowledge mapping breaks large topic areas into smaller, more manageable goals.

- Knowledge maps do not describe content.

- Knowledge maps discover correlations between goals.

Questions to Ask Yourself

- How can I split the primary goal into increasing smaller sub-goals?

- Have I accounted for every known goal within an experience?

- What is the smallest meaningful data point to record?

Kano Modeling

Chuck Noland fell out of the sky and washed up on an island. Within minutes of experiencing engine failure, his FedEx MD-11 cargo plane crashed into the dark, churning waves of the Pacific. The next morning, he awoke on the windswept shoreline of an uncharted oceanic island. Exhausted and alone, he scurried to gather together an armful of waterlogged FedEx packages that littered the beach. Each package's contents held an assortment of items, though none would reach their intended destination on time, for they were also stranded on the island, having fallen out of the plane's cargo hold the night prior. Some items would prove beneficial, while others would not—a familiar dilemma for castaways[1], as well as software users.

The packages contained the following items: a pair of Riedell Total Competition ice skates, a Wilson volleyball, a box of video cassettes, a dissolution of marriage decree, and a trashy evening dress. A pocketknife would have been helpful. So too would have a water filter or a satellite phone. However, none of those were available to Tom Hanks' character in the movie *Cast Away*. On the plus side, the island provided a wealth of natural resources, including coconuts, shellfish, and a reasonably comfortable cave. He would have to make do with the items in the packages, the island's natural resources, and nothing more.

---

[1] *Cast Away.* Directed by Robert Zemeckis. Performed by Tom Hanks. 20th Century Fox, 2000. Film.

© Edward Stull 2018
E. Stull, *UX Fundamentals for Non-UX Professionals*,
https://doi.org/10.1007/978-1-4842-3811-0_41

Software users face a similar dilemma—they must make do with what we give them. We build little islands of digital experiences, such as websites, apps, and applications. Then we supply these experiences with features. Features range from simple functionality to complex utilities. One feature may help a user log in to a website. Another may apply a photo filter within an app. Yet, we often neglect to supply them with the right features. Like island castaways collecting supplies, users will use features they need and discard the ones they do not. The features you provide determine whether their experience is a struggle or a pleasure.

Struggle or Pleasure

In the early 1980s, Noriaki Kano, professor emeritus at Tokyo University, developed a model to understand and communicate customer satisfaction. The model diagrams what we all intuitively understand, yet can never quite communicate: the value of a feature is not subjective.

Kano first used the model to describe customer satisfaction with table clocks and TV sets. Although he was not likely thinking of Tom Hanks' future film role at the time, Kano's model serves equally as well to describe the value of items on our castaway's island.

Imagine for a moment that you wish to design an island. You can provision it with a limited number of features, ranging from coconuts to satellite phones. What would you choose?

Basic features, such as food, water, and shelter, make an experience survivable. High-performance features, such as good weather, make an experience manageable. Delightful features, such as satellite phones, make an experience pleasurable. After all, what castaway would not want access to a satellite phone?

Unlike island castaways, if software users' needs are not being met, they can swim away from our island and go to someone else's. Abandonment and defection are always options.

So, the question becomes which basic, high-performance, or delightful features turn a struggle into a pleasure?

Kano's model shows the relationship between a user's satisfaction and the effort needed to achieve it (see Figure 41-1). This relationship is frequently represented on two axes: satisfaction on the y-axis, effort on the x-axis.

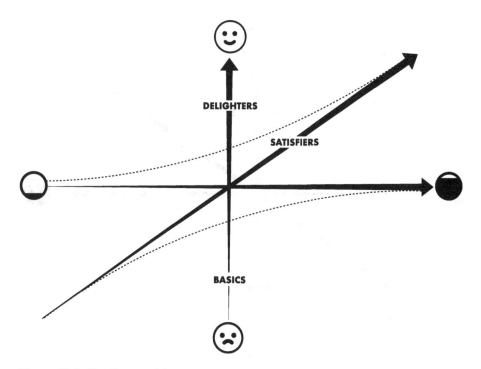

Figure 41-1. The Kano model

Let us create a model to show the satisfaction with the ice skates (see Figure 41-2). We will first need to understand how they will be used. If Tom Hanks' character used the ice skates for ice skating, they would not hold much value. Firstly, it would take a huge effort for a single man to construct an ice skating rink on a tropical island. Secondly, ice skating is an enjoyable activity, but it offers little satisfaction to a thirsty and hungry castaway.

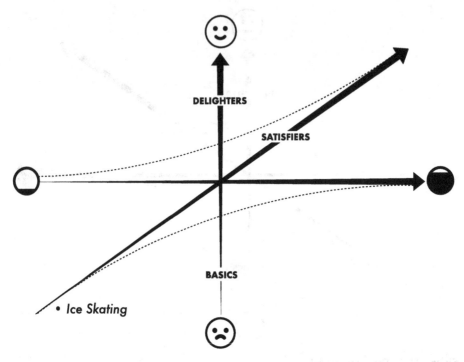

Figure 41-2. A simple Kano model

But when we change the activity from ice skating to opening coconuts, we arrive at a different model (see Figure 41-3). Using the ice skate to open a coconut requires little effort; an ice skate can easily be fashioned into an axe by holding it upside down and slamming it into an object—it is a perfect tool for opening coconuts. Considering that a castaway must drink fresh water and eat food to survive, we can safely assume that the coconut water and flesh would provide a lot of satisfaction.

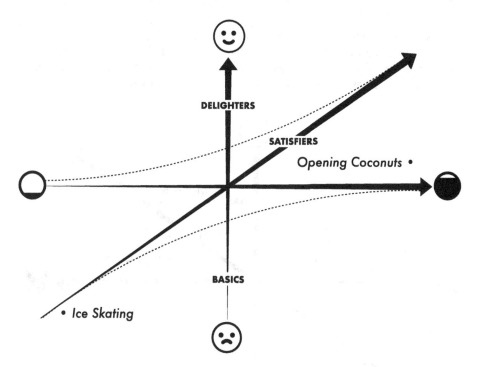

Figure 41-3. Kano model focusing on delight

Achieving delight requires more than a fully implemented, basic feature. Your island many contain a comfortable cave. You can dress it up with palm leaves, paint pictures on its walls, maybe even eventually call it home. But a comfortable cave is still a cave. Basic features rarely delight.

Achieving delight requires more than a high-performing feature. The best weather in the world only amounts to so much satisfaction when you are a castaway. Regardless of how many bright, sunny days you experience, you are still stuck on an island.

As Kano demonstrated, achieving delight is a matter of building fully implemented, high-performing features (see Figure 41-4). What would delight a visitor to the island you designed: a tall ladder to harvest coconuts; a feather mattress for the cave; or a picnic lunch to greet a castaway?

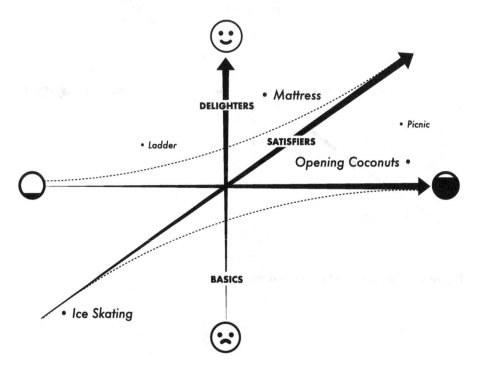

Figure 41-4. A Kano model with several delightful features

Delightful Becomes Expected Over Time

Familiarity does not breed contempt as much as it erodes it. Delight suffers the same affliction. Over time, once delightful features degrade into basic ones (see Figure 41-5). Luckily, not many of us will ever become a castaway. But, if becoming one were an everyday occurrence, we would demand more than an island's basic features. We would insist on beautiful, warm weather. We might even come to expect a picnic lunch awaiting us on the shore.

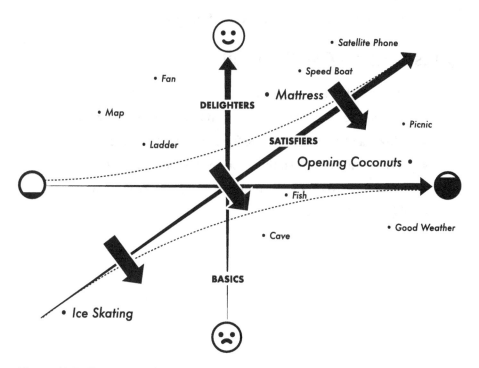

Figure 41-5. The erosion of delight

Kano addressed this erosion of delight by showing that delightful features become basic features over time. The "delight" curve dips downward and continues its descent day-by-day. Maintaining delight is a continuous effort. Many of the new websites and apps we see today will stagnate in a matter of months, if not weeks. We must continually innovate because our perception of delight is as fluid as any ocean. And, as our famous castaway once said, "Tomorrow the sun will rise; who knows what the tide will bring in."

Key Takeaways

- Kano's model shows the relationship between a user's satisfaction and the effort needed to achieve it.

- A feature's value is not subjective.

- Delightful features are fully implemented and high performing.

- Delightful features become basic features over time.

Questions to Ask Yourself

- What is the user's context?

- Is a feature useful in one context and irrelevant in another?

- How can I make a satisfying feature a delightful one?

- What features return the least on their investment?

- How will users' expectations of a feature change over time?

Heuristic Review

Every few months I cook for a group of friends. I like to cook, but my dinners are more a series of ill-timed surprises than cohesive meals. I serve food at random intervals. A main dish may be served in 20 minutes; salad, 30 minutes later; bread, after everyone finishes eating; side dishes, abandoned in the oven for hours, never make an appearance. However, with enough wine, everything generally works out in the end.

At any time during my dinner, an experienced cook could point out what needs fixing: a pinch of salt here, turn the heat up there, put out the fire in the oven, and so on. This type of review is a heuristic: an analysis and grading of individual parts. You can apply a heuristic to nearly anything, from the preparation of a meal to the user experience of software.

Some meals are better than others. Rather than say, "This dinner is terrible," a dispassionate reviewer might give it a passing or failing grade, "Dinner: Fail." Likewise, rather than say, "The chicken paprikash tastes like burnt Saran Wrap," it would be more precise to rate it as "Main Dish: Severity 5."

Applications, especially large ones, can be overwhelming to analyze in their entirety. However, large or small, applications are made of parts. We can evaluate each part and determine if it is acceptable, like a food critic evaluating each of a meal's courses.

© Edward Stull 2018
E. Stull, *UX Fundamentals for Non-UX Professionals,*
https://doi.org/10.1007/978-1-4842-3811-0_42

Heuristic scoring uses pass/fail grades, 0–5 ranges, and 0–100 percentages, or any combination of numeric or Boolean values.

We describe the pass/fail acceptability of a home page like the following:

> Home page (Pass)

Breaking down the home page into individual parts further clarifies our review:

> Search field (Pass)
>
> Hero image (Fail)
>
> Body copy (Pass)

Grading parts on a number scale (see Figure 42-1) is a more exacting approach, allowing us to compute an average score. Whereas 0 is severely problematic, 5 is marvelous:

> Search field (5)
>
> Hero image (2)
>
> Body copy (4)
>
> --------------
>
> Home page (3.6) = The average score of the page

Of note: a heuristic's average is the sum of all scores divided by the number of scores. (5 + 2 + 4) / 3 scores = 3.6

Figure 42-1. Assigning a numerical score to screen elements

In 1990, noted [1] usability experts Jakob Nielsen and Rolf Molich created a robust set of usability heuristics that are still in use today. The Nielsen heuristics cover everything from aesthetics to error prevention. Read more about "10 Usability Heuristics for User Interface Design" at https://goo.gl/zoQKAN. Researchers use several other frameworks, as well, such as Gerhardt-Powals and Weinschenk and Barker.

We can use such heuristics—or any other set you devise—to further evaluate each part of an application:

> Search field, aesthetics (4)
>
> Search field, error prevention (5)
>
> Search field, copywriting (4)

---

[1] Website Home Page with Heuristics. Digital image. Hot Sauce Market. Accessed June 7, 2018. https://hotsauce.market/.

Search field, performance (5)

Search field, HTML input type (5)

Search field (4.6)

Once you apply a heuristic across your application, you indicate what works and what does not—what needs your attention and what can wait. Perhaps you will choose to tackle parts graded from three to five, or fix everything marked as a "fail."

Evaluate each part, make improvements, and your guests will return for second helpings.

Key Takeaways

- Heuristic scoring uses numeric or Boolean values to evaluate the fitness of an experience.

- Several existing heuristic evaluations exists, such as Nielsen and Gerhardt-Powals and Weinschenk and Barker.

Questions to Ask Yourself

- What parts of an experience do I wish to evaluate?

- Should I create my own heuristics or leverage an existing framework?

- What is the most appropriate scoring method for the evaluation—pass/fail, number range, or percentage?

- How can I pair the heuristic evaluation with user testing and other research activities?

User Testing

Kobayashi Maru.

To merely mention the name Kobayashi Maru[1] invites debate among Trekkies, the devoted followers of all things *Star Trek*. It is a test—a computer simulation. Participants take the test to evaluate their leadership skills by virtually commanding a spaceship traveling across the galaxy. The Kobayashi Maru test uncovers hidden weaknesses and unforeseen strengths—a practice not unlike user testing.

Now, it is your turn. You sit in the captain's chair. Shortly after the test begins, you receive a distress call from the Kobayashi Maru, another spaceship, which sits damaged and unmovable across a contested border. The ship's crew cries out for your help. To rescue the Kobayashi Maru, you must cross the border. Yet, to do so could cause a war and lead to your own destruction.

Do you try to sneak across the border? Do you fight? Do you run? Whatever choice you make, whatever action you take, you will fail. Failing is certain, for this is how the test is designed.

The star of *Star Trek* was Captain James T. Kirk. If you are familiar with the story,[2] you know that when he faced the Kobayashi Maru, he failed it, too. But on his third attempt, he was successful. How did he win in a no-win scenario? He cheated. He reprogrammed the computer simulation to turn

---

[1] "Kobayashi Maru." Wikipedia. May 30, 2018. Accessed June 09, 2018. https://en.wikipedia.org/wiki/Kobayashi_Maru.
[2] *Star Trek II—the Wrath of Khan*. Directed by Nicholas Meyer. 1982.

© Edward Stull 2018
E. Stull, *UX Fundamentals for Non-UX Professionals*,
https://doi.org/10.1007/978-1-4842-3811-0_43

a no-win scenario into a no-lose. Kirk discovered the moments of failure within the simulation and changed them into moments of success. His actions demonstrate a vital lesson about testing: sometimes you have to lose to learn how to win.

For those who are new to user testing, the concept can sound frightening and dramatic. User testing exposes all your hard work to the whims and opinions of strangers. "What if they don't like what we built?" you wonder. "What if they hate it?"

Take a moment to imagine people testing an application you designed. Test participants flow through your application, link by link and screen by screen. You start to think, "Hey, this testing thing isn't so bad." Then it happens.

A tester clicks a link. He pauses for a moment. He clicks the back button. He tries another button. He tries again. He gets lost. He gets frustrated. You watch your design take on damage, as he shoots barrages of criticisms and vents his anger into open space. "Raise the shields," you scream. Alarms blare. Fires burn. Sparks shoot across the room as wires dangle from the rafters. Soon after, your once-promising application floats lifeless, scorched and battered, surrounded by a debris field of scribbled Post-It notes and haphazard observations.

Of course, that scenario is fictitious. Testing is far less dramatic and far more practical than many believe. More often than not, testers blame themselves for failures more than they blame the software that they are testing. They feel incapable—sometimes even stupid. They direct their frustration inward, not at you. As software creators, we should never fear testers; we should only feel empathy for them. They experience moments of failure so that we may design moments of success.

User testing strives for discovery, not destruction. We discover the hidden weaknesses and unforeseen strengths of software: the stuff we do not otherwise notice as captains of our own creations.

Rather than the high-stakes drama of the previous testing scenario, testing tends to go much more like this. You sit in a room. You greet a participant as she walks in. "Thanks for coming in today," you say. "Sure, I hope I can help," she replies. You ask her to complete a task, such as buy an airline ticket online. She does her best. You record your observations. After a few minutes, her smile transforms into pursed lips. She lets out a small, "hmm." Your ears perk up and your eyes widen, as you take notice of her mouse pointer floating across her screen. She searches and clicks. She searches and clicks again. The hmm becomes a "hmmpf!" She is lost. You wait a few seconds and ask, "How do you think you'd get back to the previous screen?" That is about as dramatic as it gets. No alarms. No fires. At most, you see a few sparks.

Qualitative and Quantitative Testing

Let's start with a testing method you can use today: qualitative testing. You can run a qualitative test at any time during a project. It's quick. It's painless. It's helpful.

Please take a moment and read the following paragraph aloud. Whisper it to yourself if you wish. Ready, go!

> Testing is quick, painless, and helpful. I'm participating in a test right now.

If you read this line aloud, we just ran a qualitative test together, albeit a small one. I asked you to do something, and you attempted to do so. Qualitative testing shares similarities with surveying. In surveying, we ask people questions. In testing, we ask people to perform tasks. Qualitative testing offers insights based on what you observe when participants perform those tasks. For example, you might ask a participant to locate information about NASA's Curiosity Mars rover mission. You notice that she first visits Google and searches for "NASA Mars." We ask her why she chose that phrase. She replies, "I recall hearing about a NASA and Mars website." She clicks the first link listed in the search results "mars.nasa.gov." After a few moments, she scrolls down the page, pauses to review it, and then finds a tout for "Looking for Curiosity?" We ask her about why she is pausing, and she tells us she was looking for the word "Curiosity."

On the surface, such a test reveals little insight; however, it may indicate the future behaviors of other users. The participant recalled hearing about a similar website, potentially signaling an audience's exposure to press coverage. She clicked the first search result, possibly demonstrating which website a future user may choose. We witnessed her pausing and scanning the page for the term "Curiosity," perhaps highlighting the importance of the term. All observations must be taken with a grain of salt, however. Qualitative tests help us understand how some users may perceive an experience, but it does not prove anything. It poses a question about each observation: "Will other people experience the same?" User testing does not provide an answer, but we should take comfort in the ubiquity of this dilemma. As the medical researcher Jonas Salk once wrote, "What people think of as the moment of discovery is really the discovery of the question."

Quantitative methods enter the equation when you score an observation. This score can be any type of quantification, but it is frequently a success/fail or numerical tally. For example, you observe 50% of participants cannot successfully find an interface button. How can we prove others will perform in a particular way?

A complete explanation of confidence intervals, error rates, samples, and populations warrants its own book, but the short answer is that truly quantitative tests require lots of participants. To reach a 95% confidence with a ±5% margin of error, we would need to test approximately 377 randomly selected people (using a population of 20,000). That is quite an effort, and often one too daunting for a typical user test. This is why most user tests tend to be purely qualitative, or a mixture of qualitative and low confidence quantitative.

We could measure the time it takes users to complete a task. Participant A takes 2 minutes. Participant B takes 3 minutes. Participant C takes 4 minutes. Afterward, we tally the results, giving us an average of 3 minutes ([2 + 3 + 4] / 3 results = 3). The more participants we add to our test, the greater the confidence of our result. You will find that similar scoring can be used to determine the average of all sorts of numerical measurements. However, always be wary of making decisions based on small sample sizes alone. Augment your tests with qualitative questions to help bolster or disprove its claims.

Remote Testing

I am a remote testing convert. The idea of remote testing seemed absurd to me at first. How can you run a test without being in the room with the participant? How could you gauge the participant's attitude, emotional state, or comfort level? Then I ran a few remote tests. I remained unconvinced until I heard a baby cry.

Many remote and online testing services record (see Figure 43-1) a test participant's computer screen. You see recordings of the tests much in the same way that the participants do, including how participants set up their home computers, their desktop backgrounds, and all the crap they leave on them. You see browser toolbars, running applications, and even the occasional indicators of viral infections. The most telling information you receive from remote testing is the audio. Not only do you hear what the participant is saying, but you also hear everything else going on around him or her. During the testing of a financial website, I heard a baby cry. The participant paused the test, came back, and I could hear the nearby baby jostling and cooing.

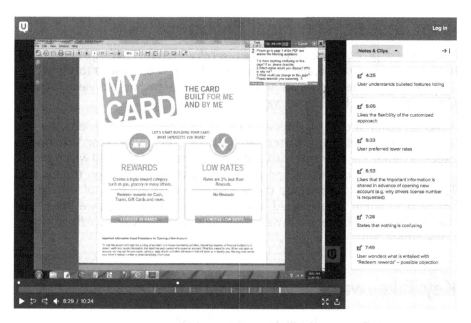

Figure 43-1. Several remote testing solutions (e.g., usertesting.com) allow facilitators to record timestamped annotations while analyzing a recording. Such annotations are helpful when evaluating why and when an event occurred during a test[3].

So, what does such a remote test tell us? For one, it tells us that whatever testing environment we set up in a laboratory will be light years ahead of what most participants have at home. It is a sobering realization that while software creators tend to have fast processors, high-resolution screens, and the latest OS updates, a sizable proportion of Americans do not. If your software is for home use, there is no better place to test software than on a participant's home computer. Perhaps most importantly, participants often feel more comfortable in their own homes than they do in a lab. They pause. They tend to their kids. They answer phone calls. They use your software in the context of their own messy lives, not in the context of your organized lab.

Still, face-to-face interaction will always have its place in user testing. With the increased need to test gestures on mobile devices, it is helpful to see both participants and what they are testing. He or she may hold her phone with one hand and swipe with the other. They may turn their tablets from portrait to landscape and rest them on their knees. Someone may be vision impaired or hard of hearing—all things perhaps best suited to test in a controlled environment. Only you can determine when face-to-face or remote testing is preferable. Regardless of which you choose, you will still find value in the discoveries, sights, and sounds revealed when testing your work.

---

[3]UserTesting. "UserTesting | User Experience Research Platform." UserTesting Blog. Accessed June 07, 2018. https://www.usertesting.com/.

User Testing: The Final Frontier

You can test prototypes, visual design, wireframes, cocktail napkin sketches, behaviors, nomenclature, and sentiment—in fact, you can test almost anything. The border between ignorance and evidence is far-reaching but easily crossed. Only one obstacle blocks our way.

Although financial cost may occasionally be a barrier to user testing, the real impediment is fear. Fear that testing displaces prerogative. Fear that testing exposes ineptitude. Fear that testing threatens creativity. None of which proves true. Prerogative, ineptitude, and creativity remain whether you test or not. Testing allows you to discover the strengths and weaknesses of software— not of the people who created it. Once we accept this fact, we lower our shields. We seek out new knowledge and new challenges and boldly go where many others will.

Key Takeaways

- Testers often blame themselves for failures.
- User testing strives for discovery, not destruction.
- User tests may not be representative of larger populations.
- Qualitative users tests do not prove anything.
- User tests tend to be purely qualitative, or a mixture of qualitative and low confidence quantitative.
- If your software is for home use, there is no better place to test software than on a participant's home computer.
- We can test almost anything.
- Testing discovers the strengths and weaknesses of software—not of the people who created it.

Questions to Ask Yourself

- Where is the best location to conduct the test?
- What hardware and software do test participants typically use?
- What tasks were completed successfully or abandoned by users?

- Where did users struggle the least and the most?

- What nonverbal cues did participants (e.g., fidgeting in chair, looking around room, or wincing at screen) make during the test?

- How long did users take to complete a task?

- What interruptions happened to the user during the test (e.g., attended to a child, dealt with a computer problem, or answered a client call)?

- How can I include my team in user testing observations?

- How can I alleviate my team's fears about user testing?

Evaluation

In 1871, Lewis Carroll wrote the novella, *Through the Looking-Glass, and What Alice Found There*, the sequel to the popular book *Alice's Adventures in Wonderland*. During Alice's travels through an alternate world made of talking flowers, twin brothers, and an anthropomorphic egg, she finds the Red Queen. The Red Queen stands the size of a full-grown adult (see Figure 44-1). Modeled after the queen in a chess set, she moves with blazing speed in any direction she wishes. She traverses the countryside, giving Alice questionable and cryptic advice. At one point in the story, she and Alice venture up a hill. Upon reaching the top, they begin to run. They run faster and faster. Yet, neither moves. Alice asks the Red Queen why. The Red Queen responds: "Now, here, you see, it takes all the running you can do to keep in the same place."

© Edward Stull 2018
E. Stull, *UX Fundamentals for Non-UX Professionals*,
https://doi.org/10.1007/978-1-4842-3811-0_44

Figure 44-1. The Red Queen and Alice from Lewis Carroll's Through The Looking-Glass and What Alice Found There. Illustration by John Tenniel, 1871[1]

A hundred years after the publication of *Alice in Wonderland*, the American biologist, Leigh Van Valen, used the exchange between Alice and the Red Queen to explain the extinction of species. The "Red Queen hypothesis"[2] describes coevolution: When two or more entities (anything ranging from species to countries to multinational corporations) compete for resources, they must

---

[1]Illustration by John Tenniel of the Red Queen lecturing Alice for Lewis Carroll's "Through The Looking-Glass" 1871, public domain.
[2]"Red Queen Hypothesis." Wikipedia. May 26, 2018. Accessed June 09, 2018. https://en.wikipedia.org/wiki/Red_Queen_hypothesis.

continually evolve to keep up with one another. They lock themselves into a continual arms race. An advantage to one causes a disadvantage to the other. This disadvantage fosters an adaptation or causes an extinction. Each entity must keep running, simply to maintain its own position and not fall behind.

Like Alice, we all strive to advance. We wish to move forward in our professional and personal lives. Experiences dominate this landscape, either helping or hindering our efforts to gain ground. Bad experiences halt progress; good ones speed us along our way. The question becomes what makes a good experience.

What Is "Good" UX?

One could argue that good user experience is *efficient* user experience: a user's time and energy are treated as precious commodities and spent only when necessary. Efficiency is certainly an important part of crafting experiences, but I bet you can think of many experiences that are worthwhile and not efficient: a good meal, a romantic vacation, a fun video game, an entertaining movie. Inefficiency is the hallmark of many good experiences.

One could argue that good user experience is a matter of *ease-of-use*: make the experience as easy as possible and it will be good. However, I bet you could think of experiences that were not necessarily easy but that you still found rewarding. Learned a new recipe? Went whitewater rafting? Beat the monster at the end of a video game? Many experiences are not easy, but you would not change them—even if given the chance.

One could argue that good user experience *delights* a user: make the experience pleasant and users will endear themselves to your creation. Yet again, we find ourselves in familiar territory. Think of an unpleasant experience in which you willfully engage: horror films, hot sauce, and hard exercise workouts, to name just a few. The experience may have been grueling; however, you willfully participate in it time and again.

If achieving good UX were only a matter of designing software with a stopwatch in one hand and a scorecard in the other, we would have perfected software design by now. eBay would have perfected auctions. Amazon would have perfected e-commerce. Facebook would have perfected social media. And every other digital archetype would exist in an ideal state of timeless perfection. But we know this to be untrue. Although these applications are successful, they constantly change—they pivot and tweak. New competitive companies enter the market. New devices come out. New attitudes emerge. New patterns unfold.

What we are left with is a problem, but a clearly stated one: the potential for both good and bad UX is built into every product, service, function, interaction, and piece of content. It is not one thing that makes an experience succeed or fail. It is everything. A good experience attempts to solve this unsolvable riddle. It makes the effort. It races up the hill, striving to advance, if only to keep its users from falling behind.

Key Takeaways

- UX is in a constant state of change.
- The potential for both good and bad UX is built into every product, service, function, interaction, and piece of content.
- No one thing makes an experience succeed or fail.
- Good UX serves users.

Questions to Ask Yourself

- At the minimum, does an experience preserve user safety, security, and dignity?
- Does any part of an experience hinder users' efforts to complete their goals?
- Am I making the effort to provide users with a good experience?

Conclusion

In this book, we met a French lieutenant, a manicurist, and a Mexican dog.

We sat on a riverbank in Sri Lanka and talked about American woodpeckers, German fly catchers, and Russian folklore. World War II tanks rolled past us. We rode the Hiawatha, witnessed UFOs, and slept in the bed of a Greek psychopath. We counted bones.

Irradiated Missourians persuaded us. We learned about ethos, pathos, and bears that run as quickly as house cats. Authority gave way to relevancy in the flowering hills of Munnar. Dorothy Parker showed us we could all laugh, even when destroying a laser printer.

Pandas and baby oxen brought us gifts. King Frederick grew us potatoes. Tolstoy and Myspace came at a cost. Chinese rail workers and philosophers taught us about marketing.

North Korea helped us run a project. Waterfall, Agile, and Lean dug us a tunnel. Along the way, we played *The Dating Game*, got tattoos, and drank gunpowder and rum.

Star Trek's Kobayashi Maru flew by Plato's cave and left us on an tropical island. The Red Queen raced up a hillside. From there, we charted our journey to the end.

The end of any journey elicits at least one disclaimer—every project, client, and team is different. What works well in one place may fall flat in another. This book contains a wealth of opinions, advice, and observations; but they all pale in comparison to your unique understanding of your own circumstances.

Although this book's knowledge is limited, its lessons are universal. Sometimes you are the creator; sometimes you are the user. Sometimes you are the author; sometimes you are the reader. But what remains is always an experience. I hope yours was a pleasant one.

© Edward Stull 2018
E. Stull, *UX Fundamentals for Non-UX Professionals*,
https://doi.org/10.1007/978-1-4842-3811-0_45

Resources for Further Reading

You can find a wealth of additional information about UX—and the book's other topics—by reviewing the following list of contributing resources.

Introduction

Bristow, Michael. "China's Internet 'spin Doctors'." BBC News. December 16, 2008. Accessed May 28, 2018. http://news.bbc.co.uk/2/hi/7783640.stm.

Holloway, April. "The Toraja People and the Most Complex Funeral Rituals in the World." Ancient Origins. Accessed May 28, 2018. http://www.ancient-origins.net/ancient-places-asia/toraja-people-and-most-complex-funeral-rituals-world-001268.

"Traditions - Girl Scouts." Girl Scouts of the USA. Accessed May 28, 2018. http://www.girlscouts.org/program/basics/traditions/swaps/.

Part I: UX Principles

Cartwright, Mark. "Hercules." Ancient History Encyclopedia. May 29, 2018. Accessed May 29, 2018. https://www.ancient.eu/hercules/.

"Hercules' Second Labor: The Lernean Hydra." The Perseus Project. Accessed June 07, 2018. http://www.perseus.tufts.edu/Herakles/hydra.html.

© Edward Stull 2018
E. Stull, *UX Fundamentals for Non-UX Professionals*,
https://doi.org/10.1007/978-1-4842-3811-0

Chapter 1: UX Is Unavoidable

O'Barr, William M. "A Brief History of Advertising in America." Advertising & Society Review. September 01, 2005. Accessed May 28, 2018. `http://muse.jhu.edu/journals/asr/v006/6.3unit02.html`.

Chapter 2: You Are Not the User

Planet, Lonely. *Lonely Planet Discover China.* Lonely Planet, 2017.

Chapter 3: You Compete with Everything

Levy, Jaime. UX Strategy: *How to Devise Innovative Digital Products That People Want.* Sebastopol, CA: O'Reilly, 2015.

Lovett, Charles C. *Olympic Marathon: A Centennial History of the Games' Most Storied Race.* Westport, CT: Praeger, 1997.

Chapter 4: The User Is on a Journey

"Le Marathon Du Médoc - Site Officiel." Le Marathon Du Médoc - Site Officiel. Accessed May 28, 2018. `http://www.marathondumedoc.com/`.

"Marathon Du Medoc a Marathon Drinking Session." *Runner's World.* October 09, 2008. Accessed May 28, 2018. `http://www.runnersworld.co.uk/event-editorial/marathon-du-medoc-a-marathon-drinking-session/3716.html`.

"Marathon Men." International Olympic Committee. June 03, 2017. Accessed May 28, 2018. `http://www.olympic.org/olympic-results/london-2012/athletics/marathon-m`.

Chapter 5: Keep It Simple

Buckley, John D. *British Armour in the Normandy Campaign, 1944.* London: Cass, 2004.

Jentz, Thomas L. *Germany's Tiger Tanks: Tiger I & Tiger II: Combat Tactics.* Atglen, PA: Schiffer Publishing, 1997.

Richmond, Shane. "Call of Duty: Modern Warfare 3 Breaks Sales Records." The Telegraph. November 11, 2011. Accessed May 28, 2018. `http://www.telegraph.co.uk/technology/video-games/video-game-news/8884726/Call-of-Duty-Modern-Warfare-3-breaks-sales-records.html#`.

Chapter 6: Users Collect Experiences

"Katamari Damacy." Katamari Wiki. Accessed May 28, 2018. http:// katamari.wikia.com/wiki/Katamari_Damacy.

Chapter 7: Speak the User's Language

"The Rosetta Stone." British Museum. Accessed May 28, 2018. http:// www.britishmuseum.org/research/collection_online/collection_ object_details.aspx?objectId=117631&partId=1&searchText=rosett a%2Bstone&page=1.

Saunders, Nicholas J. *Alexander's Tomb: The Two Thousand Year Obsession to Find the Lost Conqueror.* New York: Basic Books, 2007.

Chapter 8: Favor the Familiar

Gibson, James J. *The Ecological Approach to Visual Perception.* New York: Psychology Press, 2015.

Norman, Donald A. *The Design of Everyday Things.* New York: Basic Books, a Member of the Perseus Books Group, 2013.

Chapter 9: Stability, Reliability, and Security

Betts, Gwendolyn. "Security Vs. UX: How To Reconcile One Of The Biggest Challenges In Interface Design." Co.Design. May 02, 2017. Accessed May 28, 2018. http://www.fastcodesign.com/3059293/security-vs-ux-how-to-reconcile-one-of-the-biggest-challenges-in-interface-design.

"Kevin Mitnick Answers - Slashdot." Http://slashdot.org/. Accessed May 28, 2018. http://slashdot.org/story/03/02/04/2233250/Kevin-Mitnick-Answers.

PBS. Accessed May 28, 2018. http://www.pbs.org/wgbh/pages/frontline/ shows/hackers/whoare/notable.html.

Stefanov, Stoyan. "Book of Speed." Book of Speed. 2011. Accessed May 28, 2018. http://www.bookofspeed.com/.

"Transatlantic Optical Cable." Transatlantic Optical Cable - Engineering and Technology History Wiki. Accessed May 28, 2018. http://ethw.org/ Transatlantic_Optical_Cable.

Chapter 10: Speed

"Firefox & Page Load Speed – Part II." Blog of Metrics. Accessed May 28, 2018. https://blog.mozilla.org/metrics/2010/04/05/firefox-page-load-speed-%E2%80%93-part-ii/.

Linden, Greg. "Geeking with Greg." Marissa Mayer at Web 2.0. January 01, 1970. Accessed May 28, 2018. http://glinden.blogspot.com/2006/11/marissa-mayer-at-web-20.html.

Stoyan Stefanov, Engineer Follow. "YSlow 2.0." LinkedIn SlideShare. December 11, 2008. Accessed May 28, 2018. https://www.slideshare.net/stoyan/yslow-20-presentation.

Chapter 11: Usefulness

"The Latest Thing". *Uncle John's Legendary Lost Bathroom Reader*. Portable Press. p. 373. ISBN 1-879682-74-5.

Part II: Being Human

"10 Weird Religious Practices." Listverse. June 25, 2014. Accessed May 28, 2018. http://listverse.com/2007/08/13/10-weird-religious-practices/.

"Births and Natality." Centers for Disease Control and Prevention. March 31, 2017. Accessed May 28, 2018. http://www.cdc.gov/nchs/fastats/births.htm.

"Birthing Rituals: The Weird, the Wacky, and the Truly Bizarre." *The Sydney Morning Herald*. October 03, 2011. Accessed May 28, 2018. http://www.smh.com.au/lifestyle/life/blogs/dirty-laundry/birthing-rituals-the-weird-the-wacky-and-the-truly-bizarre-20111004-1l5n1.html.

"Celestis: Memorial Spaceflights - Space Funeral Ashes Burial." Celestis: Memorial Spaceflights - Space Funeral Ashes Burial. Accessed May 28, 2018. http://www.memorialspaceflights.com/.

Moskowitz, Clara (2012-05-22). "Ashes of Star Trek's 'Scotty' Ride Private Rocket Into Space". Space.com. New York. Archived from the original on 2012-05-22. Retrieved 2018-04-04.

"Nose, Sinuses and Smell." *InnerBody*. Accessed May 28, 2018. http://www.innerbody.com/anim/nasal.html.

"News." Trends in Funeral Service. Accessed May 28, 2018. http://nfda.org/about-funeral-service-/trends-and-statistics.html.

Spector, Dina. "INFOGRAPHIC: Amazing Facts About The Human Body." *Business Insider*. August 01, 2011. Accessed May 28, 2018. http://www.businessinsider.com/infographic-human-body-2011-8.

Chapter 13: Perception

Assael, Henry. *Consumer Behavior and Marketing Action*. Noida: Thomson, 2006.

Controlled and Automatic Processing. Accessed May 28, 2018. http://www.indiana.edu/~p1013447/dictionary/ctrlauto.htm.

Efran, Michael G. "The Effect of Physical Appearance on the Judgment of Guilt, Interpersonal Attraction, and Severity of Recommended Punishment in a Simulated Jury Task." *Journal of Research in Personality* 8, no. 1 (1974): 45-54. doi:10.1016/0092-6566(74)90044-0.

"Gestalt Principles of Form Perception." The Interaction Design Foundation. Accessed May 28, 2018. http://www.interaction-design.org/encyclopedia/gestalt_principles_of_form_perception.html.

Haidt, Jonathan. *The Righteous Mind: Why Good People Are Divided by Politics and Religion*. New York: Pantheon Books, 2012.

Hoagland, Richard C. *The Monuments of Mars: A City on the Edge of Forever*. Berkeley, CA: Frog, 2002.

"How Likely Are You to Live to 100? Get the Full Data." *The Guardian*. August 04, 2011. Accessed May 28, 2018. http://www.theguardian.com/news/datablog/2011/aug/04/live-to-100-likely.

Jiang, J., B. Dai, D. Peng, C. Zhu, L. Liu, and C. Lu. "Neural Synchronization during Face-to-face Communication." The Journal of Neuroscience: *The Official Journal of the Society for Neuroscience*. November 07, 2012. Accessed May 28, 2018. http://www.ncbi.nlm.nih.gov/pubmed/23136442.

Paradise Lost: Overview. Accessed May 28, 2018. http://www.paradiselost.org/5-overview.html.

Pinker, Steven. *The Better Angels of Our Nature: A History of Violence and Humanity*. London: Penguin Books, 2012.

Thorndike, E.I. "A Constant Error in Psychological Ratings." *Journal of Applied Psychology* 4, no. 1 (1920): 25-29. doi:10.1037/h0071663.

Goldstein, E. Bruce and James R. Brockmole. *Sensation and Perception*. Boston, MA: Cengage Learning, 2017.

Hadjikhani, Nouchine, Kestutis Kveraga, Paulami Naik, and Seppo P. Ahlfors. "Early (M170) Activation of Face-specific Cortex by Face-like Objects." *NeuroReport* 20, no. 4 (2009): 403-07. doi:10.1097/wnr.0b013e328325a8e1.

Hoagland, Richard C. *The Monuments of Mars: A City on the Edge of Forever.* Berkeley, CA: Frog, 2002.

Weinschenk, Susan. *100 Things Every Designer Needs to Know About People.* Berkeley: New Riders, 2011.

YaleCourses. "9. Paradise Lost, Book I." YouTube. November 21, 2008. Accessed May 28, 2018. https://www.youtube.com/watch?v=H62G9yIN5Wk&list=PL 2103FD9F9D0615B7.

Chapter 14: Attention

Applied Cognitive Psychology. Chichester: Wiley, 1982.

"Attention Span Statistics." Statistic Brain. August 12, 2016. Accessed May 28, 2018. http://www.statisticbrain.com/attention-span-statistics/.

Gorayska, Barbara and Jacob L. Mey. *Cognition and Technology: Co-existence, Convergence, and Co-evolution.* Amsterdam: Benjamins, 2004.

Profsimons. "Selective Attention Test." YouTube. March 10, 2010. Accessed May 28, 2018. http://www.youtube.com/watch?v=vJG698U2Mvo.

"The Reading Brain." The Frontal Cortex. Posted by Jonah Lehrer on November 20, 2009. Accessed May 28, 2018. http://scienceblogs.com/cortex/2009/11/20/the-reading-brain-1/.

Resources, Ohio Department of Natural. "Ohio.gov/Search." Ohio Department of Natural Resources. Accessed May 28, 2018. http://ohiodnr.com/forestry/trees/pawpaw/tabid/5404/Default.aspx.

Shirky, Clay. *Cognitive Surplus: Creativity and Generosity in a Connected Age.* East Rutherford: Penguin Press, 2012.

Sternberg, Robert J., Karin Sternberg, and Jeffery Scott Mio. *Cognitive Psychology.* Australia: Cengage Learning, 2017.

Chapter 15: Flow

Csikszentmihalyi, Mihaly. *Flow: The Psychology of Optimal Experience.* New York: Harper Row, 2009.

Chapter 16: Laziness

Kahneman, Daniel. *Thinking, Fast and Slow.* New York: Farrar, Straus and Giroux, 2015.

Schindler, Melissa. "How Fast Can a Domestic Cat Run?" Back. Accessed May 28, 2018. `http://pets.thenest.com/fast-can-domestic-cat-run-10393.html`.

Chapter 17: Memory

Bernstein, Douglas A., Julie Ann Pooley, Lynne Cohen, Bethanie Gouldthorp, Stephen C. Provost, and Jacquelyn Cranney. *Psychology.* South Melbourne, Vic.: Cengage, 2017.

Brady, T. F., T. Konkle, and G. A. Alvarez. "A Review of Visual Memory Capacity: Beyond Individual Items and Toward Structured Representations." *Journal of Vision* 11, no. 5 (2011): 4. doi:10.1167/11.5.4.

Cowan, Nelson. "The Magical Mystery Four: How Is Working Memory Capacity Limited, and Why?" Current Directions in Psychological Science. February 01, 2010. Accessed May 28, 2018. `http://www.ncbi.nlm.nih.gov/pmc/articles/PMC2864034/`.

Cowan, Nelson. "The Magical Number 4 in Short-term Memory: A Reconsideration of Mental Storage Capacity | Behavioral and Brain Sciences." Cambridge Core. October 30, 2001. Accessed May 28, 2018. `http://journals.cambridge.org/action/displayAbstract?fromPage=online&aid=84441`.

Duration of Short-Term Memory. Accessed May 28, 2018. `http://www.indiana.edu/~p1013447/dictionary/stmpp.htm`.

"How Quickly We Forget: The Transience of Memory." *PsyBlog.* October 16, 2016. Accessed May 28, 2018. `http://www.spring.org.uk/2008/01/how-quickly-we-forget-transience-of.php`.

Loftus, Elizabeth. "How Reliable Is Your Memory?" Elizabeth Loftus: How Reliable Is Your Memory? | TED Talk | TED.com. Accessed May 28, 2018. `http://www.ted.com/talks/elizabeth_loftus_the_fiction_of_memory.html`.

"Peterson and Peterson (1959)." Peterson and Peterson 1959 | Simply Psychology. January 01, 1970. 2008, Saul McLeod Published. Accessed May 28, 2018. `http://www.simplypsychology.org/peterson-peterson.html`.

Magill, Richard A. *Memory and Control of Action.* Amsterdam: North-Holland, 1983.

McEntarffer, Robert and Allyson Weseley. *Barron's AP Psychology.* Hauppauge, NY: Barron's, 2016.

Chapter 18: Rationalization

http://www.esquire.com/news-politics/a4642/the-last-meal-0598/

Chapter 19: Accessibility

"Manual on Uniform Traffic Control Devices." Manual on Uniform Traffic Control Devices (MUTCD) - FHWA. Accessed May 28, 2018. http://mutcd.fhwa.dot.gov/.

Chapter 20: Storytelling

Fogg, Brian J. *Persuasive Technology: Using Computers to Change What We Think and Do.* Amsterdam: Morgan Kaufmann, 2011.

Nordquist, Richard. "John Quincy Adams on the Art of Persuasion." ThoughtCo. Accessed October 2017. https://www.thoughtco.com/art-of-persuasion-john-quincy-adams-1690757.

Pew, Richard W. and Van Hemel Susan B. *Technology for Adaptive Aging.* Washington, D.C.: National Academies Press, 2004.

Redish, Janice. *Letting Go of the Words.* Waltham, MA: Morgan Kaufmann, 2014.

Part III: Persuasion

"Enhydris Chinensis." Enhydris Chinensis (Chinese Mud Snake, Chinese Rice Paddy Snake, Chinese Water Snake, Tang Water Snake). Accessed May 28, 2018. http://www.iucnredlist.org/details/176674/0.

Dooley, Roger. *Brainfluence: 100 Ways to Persuade and Convince Consumers with Neuromarketing.* Hoboken, NJ: John Wiley, 2012.

Fogg, B.J. *Persuasive Technology Using Computers to Change What We Think and Do.* Amsterdam: Morgan Kaufmann Publishers, an Imprint of Elsevier Science, 2011.

Gregory, Richard. *The Oxford Companion to the Mind.* Oxford: Oxford University Press, 2005.

GuÃcguen, Nicolas. "Foot-in-the-door Technique and Computer-mediated Communication." *Computers in Human Behavior* 18, no. 1 (2002): 11-15. doi:10.1016/s0747-5632(01)00033-4.

National Archives and Records Administration. Accessed May 28, 2018. http://www.archives.gov/exhibits/treasures_of_congress/text/page15_text.html.

Rumelt, Richard. *Good Strategy/Bad Strategy: The Difference and Why It Matters.* S.l.: Profile Books, 2017.

Scott, David Meerman. *The New Rules of Marketing & PR: How to Use News Releases, Blogs, Podcasting, Viral Marketing & Online Media to Reach Buyers Directly.* Hoboken: Wiley, 2009.

"Snake Oil." Princeton University. Accessed May 28, 2018. http://www.princeton.edu/~achaney/tmve/wiki100k/docs/Snake_oil.html.

Chapter 21: Empathy

Bunch, Will. "Tearing Down the Reagan Myth: Now More Than Ever." *The Huffington Post.* January 31, 2010. Accessed May 28, 2018. http://www.huffingtonpost.com/will-bunch/tearing-down-the-reagan-m_b_443914.html.

Buonomano, Dean. *Brain Bugs: How the Brain's Flaws Shape Our Lives.* New York: W.W. Norton, 2012.

Copernicus, Nicolaus. *De Revolutionibus.* Berlin: Walter De Gruyter, 1973.

Diggins, John P. *Ronald Reagan: Fate, Freedom, and the Making of History.* New York: W.W. Norton & Company, 2008.

Eiseley, Loren C. *The Star Thrower.* New York: Times Books, 1978.

Hersey, John. *Hiroshima.* London: Penguin Books, 2015.

Rawls, John. *A Theory of Justice.* New Delhi: Universal Law Publishing Co, 2013.

Schell, Jonathan. *The Fate of the Earth.* New York: Knopf, 1988.

Chapter 22: Authority

Blass, Thomas. "The Milgram Paradigm After 35 Years: Some Things We Now Know About Obedience to Authority." *Journal of Applied Social Psychology* 29, no. 5 (1999): 955-78. doi:10.1111/j.1559-1816.1999.tb00134.x.

Brocas, Isabelle and Juan D. Carrillo. *The Psychology of Economic Decisions. Rationality and Well-being.* Oxford: Oxford University Press, 2003.

Milgram, Stanley. *Obedience to Authority: An Experimental View.* New York: Harper Torchbooks, 1969.

Chapter 23: Motivation

Cacioppo, John T. and Richard E. Petty. "The Elaboration Likelihood Model of Persuasion." *ACR North American Advances*. January 01, 1984. Accessed August 05, 2017. http://acrwebsite.org/volumes/6329/volumes/v11/NA-11.

Petty, Richard E. *Communication and Persuasion: Central and Peripheral Routes to Attitude Change*. Place of Publication Not Identified: Springer, 2012.

Chapter 24: Relevancy

Grossberg, Stephen. *The Adaptive Brain*. Amsterdam: North-Holland, 1992.

"Legal - ICloud - Apple." Apple Legal. Accessed May 28, 2018. http://www.apple.com/legal/internet-services/icloud/en/terms.html.

"Munnar - the Hill Station of Kerala in Idukki." Kerala Tourism. Accessed May 28, 2018. https://www.keralatourism.org/destination/munnar/202/.

"NeelakurinjiFlowers,Munnar." KeralaTourismAccessedMay28,2018.https://www.keralatourism.org/destination/neelakurinji-munnar/366.

PBS. Accessed May 28, 2018. http://www.pbs.org/parents/child-developmenttracker/one/language.html.

"Wenger 16999 Swiss Army Knife Giant - Most Expensive Item On Amazon." Most Expensive Item On Amazon - Amazon.com. Accessed May 28, 2018. https://www.amazon.com/Wenger-16999-Swiss-Knife-Giant/dp/B001D ZTJRQ#productDetails.

Chapter 25: Reciprocity

"Arkansas." The Ivory-Billed Woodpecker | *The Nature Conservancy*. Accessed May 28, 2018. http://www.nature.org/ourinitiatives/regions/northamerica/unitedstates/arkansas/ivorybill/.

Besst, Nancy. *Milton and Matilda: The Musk Oxen Who Went to China*. San Francisco: China Books, 1982.

Mauss, Marcel. *The Gift: The Form and Reason for Exchange in Archaic Societies*. London: Routledge, 1990.

Nicholls, Henry. *The Way of the Panda: The Curious History of China's Political Animal*. New York: Pegasus Books, 2011.

Sahlins, Marshall. *Stone Age Economics*. London: Tavistock, 1978.

Stec, Carly. "20 Display Advertising Stats That Demonstrate Digital Advertising's Evolution." HubSpot. Accessed May 28, 2018. http://blog.hubspot.com/marketing/horrifying-display-advertising-stats.

Chapter 26: Product

Illustration by John James Audubon, Victoria and Albert Museum/Getty Images. "Ivory-Billed Woodpecker." National Geographic. August 01, 2017. Accessed May 28, 2018. `http://animals.nationalgeographic.com/animals/birds/ivory-billed-woodpecker/`.

Carroll, Lewis. *Alice's Adventures in Wonderland*. London: Ward, Lock &, 1952.

Eyal, Nir and Ryan Hoover. *Hooked How to Build Habit-forming Products*. Princeton, NJ: Princeton University Press, 2014.

Kotler, Armstrong. *Principles Of Marketing*, 11th Edition. Place of Publication Not Identified: Academic Internet Publisher, 2006.

Chapter 27: Price

The Bridge on the River Kwai. Produced by Sam Spiegel, Donald M. Ashton, Geoffrey Drake, Cecil F. Ford, Stewart Freeborn, George Partleton, and John Wilson-Apperson. By Michael Wilson, Carl Foreman, Peter Taylor, Winston Ryder, Jack Hildyard, and Malcolm Arnold. Directed by David Lean. Performed by William Holden, Alec Guinness, Jack Hawkins, Sesshu Hayakawa, James Donald, Geoffrey Horne, André Morell, Percy Herbert, Harold Goodwin, Henry Okawa, Keiichiro Katsumoto, and M. R. B. Chakrabandhu. United States: Columbia Pictures Corp. Presents, 1957.

Cialdini, Robert B. *Influence: Science and Practice*. Harlow: Pearson Education, 2014.

Improbable Research. Accessed May 28, 2018. `http://www.improbable.com/airchives/paperair/volume9/v9i2/mankiw.html`.

Johannes, James D., Rodney D. Stewart, and Richard M. Wyskida. *Cost Estimator's Reference Manual*. New York, NY: Wiley, 1995.

Kahneman, Daniel and Amos Tversky. "Prospect Theory: An Analysis of Decision under Risk." *Econometrica* 47, no. 2 (1979): 263. doi:10.2307/1914185.

Loewenstein, G., T. O'Donoghue and M. Rabin. "Projection Bias in Predicting Future Utility." *The Quarterly Journal of Economics* 118, no. 4 (2003): 1209-248. doi:10.1162/003355303322552784.

"Oldest Performing Ballerina." Guinness World Records. Accessed May 28, 2018. `http://www.guinnessworldrecords.com/records-2000/oldest-performing-ballerina/`.

Oyedokun, Godwin Emmanuel. "Accounting Theory: Review of Theory in Purchasing and Inventory Management." *SSRN Electronic Journal*. doi:10.2139/ssrn.2912269.

Thaler, Richard. "Transaction Utility Theory." *ACR North American Advances.* January 01, 1983. Accessed May 28, 2018. http://www.acrwebsite.org/search/view-conference-proceedings.aspx?Id=6118.

Tolstoy, Leo. *The Story of Ivan the Fool, and His Two Brothers: Simeon the Warrior and Taras the Stout, and of His Dumb Sister Malania the Spinster, and of the Old Devil and the Three Devilkins.* London: Brotherhood Pub., 1898.

Shor, Mike. "Dictionary." Pareto Efficient - Game Theory .net. Accessed May 28, 2018. http://www.gametheory.net/dictionary/ParetoEfficient.html.

Chapter 28: Promotion

Kiple, Kenneth F. and Kriemhild Coneè. Ornelas. *The Cambridge World History of Food.* Cambridge, U.K.: Cambridge University Press, 2001.

"The Legend of the Potato King." *The New York Times.* October 11, 2012. https://niemann.blogs.nytimes.com/2012/10/11/the-legend-of-the-potato-king/.

Mcneill, William H. "The History and Social Influence of the Potato. Redcliffe N. Salaman , W. G. Burton." *The Journal of Modern History* 22, no. 4 (1950): 366-67. doi:10.1086/237372.

Norden, Bryan Van. "Mencius." *Stanford Encyclopedia of Philosophy.* October 16, 2004. Accessed May 28, 2018. http://plato.stanford.edu/entries/mencius/.

Rice Culture of China. Accessed May 28, 2018. http://www.china.org.cn/english/2002/Oct/44854.htm.

Snyder, Sarah. "Translation, Mencius "Pulling up Sprouts"." Academia. edu. Accessed May 28, 2018. http://www.academia.edu/5959792/Translation_Mencius_Pulling_up_Sprouts_.

Wright, Rita P. *The Ancient Indus: Urbanism, Economy, and Society.* New York: Cambridge University Press, 2010.

Chapter 29: Place

Heath, Chip and Dan Heath. *Made to Stick: Why Some Ideas Take Hold and Others Come Unstuck.* New York: Random House Books, 2010.

"The Kuleshov Experiment." The Kuleshov Experiment | Basics of Film Editing. Accessed May 28, 2018. http://www.elementsofcinema.com/editing/kuleshov-effect.html.

Thaler, Richard. "Toward a Positive Theory of Consumer Choice." *Journal of Economic Behavior & Organization* 1, no. 1 (1980): 39-60. doi:10.1016/0167-2681(80)90051-7.

Underhill, Paco. *Why We Buy?: The Science of Shopping.* New York: Simon & Schuster, 2009.

Part IV: Process

Gray, Dave. *Gamestorming.* O'Reilly Verlag, 2013.

"Pay & Leave Pay Administration." U.S. Office of Personnel Management. Accessed May 28, 2018. http://www.opm.gov/policy-data-oversight/pay-leave/pay-administration/fact-sheets/computing-hourly-rates-of-pay-using-the-2087-hour-divisor/.

Sik, Kim Hyun. "The Secret History of Kim Jong II." Foreign Policy. October 06, 2009. Accessed May 28, 2018. http://www.foreignpolicy.com/articles/2008/08/12/the_secret_history_of_kim_jong_il.

Stickdorn, Marc and Jakob Schneider. *This Is Service Design Thinking: Basics-tools-cases.* Amsterdam, The Netherlands: BIS Publishers, 2011.

"The World Factbook: Korea, North." Central Intelligence Agency. July 24, 2017. Accessed May 28, 2018. https://www.cia.gov/library/publications/the-world-factbook/geos/kn.html.

Chapter 30: Waterfall, Agile, and Lean

Gilmore, David J., Russel Winder and Francĺ§oise Deĺ tienne. *User-centred Requirements for Software Engineering Environments.* Berlin: Springer, 1994.

Gothelf, Jeff and Joshua Seiden. *Lean UX: Designing Great Products with Agile Teams.* Beijing; Boston; Farnham; Sebastopol; Tokyo: O'Reilly, 2016.

Hartson, Rex and Pardha S. Pyla. *The UX Book: Process and Guidelines for Ensuring a Quality User Experience.* Amsterdam: Morgan Kaufmann, 2016.

Chapter 31: Problem Statements

"Hutzler 571 Banana Slicer: Kitchen & Dining." Amazon.com: Hutzler 571 Banana Slicer: Kitchen & Dining. Accessed August 05, 2017. https://www.amazon.com/Hutzler-571-Banana-Slicer/dp/B0047E0EII/.

Chapter 33: Quantitative Research

"Alcoholic Proof." Princeton University. Accessed May 28, 2018. http://www.princeton.edu/~achaney/tmve/wiki100k/docs/Alcoholic_proof.html.

Covert, Abby and Nicole Fenton. *How to Make Sense of Any Mess.* S.l.: S.n., 2014.

Portigal, Steve. *Interviewing Users: How to Uncover Compelling Insights.* Sebastopol: Rosenfeld Media, 2013.

"Procrustes." Princeton University. Accessed May 28, 2018. https://www.princeton.edu/~achaney/tmve/wiki100k/docs/Procrustes.html.

Chapter 34: Calculator Research

Calabrese, Erin and Josh Saul. "That's Rich! McDonald's Tells Workers What to Tip Au Pairs." *New York Post.* December 07, 2013. Accessed May 28, 2018. http://nypost.com/2013/12/07/rich-with-irony-mcdonalds-gives-workers-advice-on-tipping-pool-boys-au-pairs/.

"Summary." U.S. Bureau of Labor Statistics. Accessed June 4, 2018. https://www.bls.gov/ooh/food-preparation-and-serving/food-preparation-workers.htm.

Chapter 35: Qualitative Research

Ekman, Paul. *Telling Lies: Clues to Deceit in the Marketplace, Politics, and Marriage.* New York: W.W. Norton, 2009.

Fitzgerald, F. Scott. *The Great Gatsby.* Oxford: Benediction Classics, 2016.

Hall, Erika. *Just Enough Research.* NY, NY: Book Apart, 2013.

Krupnick, Ellie. "Nike Tattoo Leggings Pulled After Deemed Exploitative Of Samoan Culture (PHOTOS)." *The Huffington Post.* August 15, 2013. Accessed May 28, 2018. http://www.huffingtonpost.com/2013/08/15/nike-tattoo-leggings_n_3763591.html.

Lee, Harper. *To Kill a Mockingbird.* London: Arrow Books, 2015.

"National Center for Health Statistics." Centers for Disease Control and Prevention. Accessed May 28, 2018. http://www.cdc.gov/nchs/fastats/bodymeas.htm.

Patterson, Kerry, Joseph Grenny, Al Switzler, and Ron McMillan. *Crucial Conversations: Tools for Talking When the Stakes Are High.* New York: McGraw-Hill, 2012.

Chapter 36: Reconciliation

Clark, Roy Peter. *Help! for Writers: 210 Solutions to the Problems Every Writer Faces.* New York: Little, Brown, 2013.

Club, American Kennel. "Xoloitzcuintli Dog Breed Information." American Kennel Club. Accessed May 28, 2018. http://www.akc.org/dog-breeds/xoloitzcuintli/.

"Pets by the Numbers." The Humane Society of the United States. Accessed May 28, 2018. http://www.humanesociety.org/issues/pet_overpopulation/facts/pet_ownership_statistics.html.

Roth, David Lee. *Crazy from the Heat.* London: Ebury, 2000.

Sutherland, Rory. "Sweat the Small Stuff." Rory Sutherland: Sweat the Small Stuff | TED Talk | TED.com. Accessed May 28, 2018. https://www.ted.com/talks/rory_sutherland_sweat_the_small_stuff.

Chapter 37: Documentation

Cherry, Kendra. "What Makes Human Factors Psychology Different?" Verywell. Accessed May 28, 2018. https://www.verywell.com/what-is-human-factors-psychology-2794905.

"The Darby Cooking Pot That Changed the World." BBC. January 28, 2010. Accessed May 28, 2018. http://news.bbc.co.uk/local/shropshire/hi/people_and_places/history/newsid_8485000/8485234.stm.

Goldstein, Noah J., Steve J. Martin, and Robert B. Cialdini. *Yes! 50 Scientifically Proven Ways to Be Persuasive.* New York, NY: Free Press, 2010.

Meltailiel, Georges, Dieter Kuhn, Joseph Needham, and Tsuen-hsuin Tsien. *Science and Civilisation in China.* Cambridge: Cambridge Univ. Press, 2015.

Sebastian Anthony on November 12, 2013 at 6:00 Am Comment. "How Long Do Hard Drives Actually Live For?" ExtremeTech. November 11, 2013. Accessed May 28, 2018. http://www.extremetech.com/computing/170748-how-long-do-hard-drives-actually-live-for.

Chapter 38: Personas

Rowe, Peter G. *Design Thinking.* Cambridge, Mass.: MIT Press, 1998.

Chapter 39: Journey Mapping

Hancox, David, Clive H. Schofield, and John Robert Victor. Prescott. *A Geographical Description of the Spratly Islands and an Account of Hydrographic Surveys amongst Those Islands.* Durham: International Boundaries Research Unit - Univ. of Durham, 1995.

Patton, Jeff. *User Story Mapping:.* Beijing: O'Reilly, 2014.

Chapter 41: Kano Modeling

"International Decade for Action, Water for Life," 2015, UN-Water, United Nations, MDG, Water, Sanitation, Financing, Gender, IWRM, Human Right, Transboundary, Cities, Quality, Food Security, General Comment, BKM, Albuquerque." United Nations. Accessed May 28, 2018. http://www.un.org/waterforlifedecade/human_right_to_water.shtml.

Chapter 42: Heuristic Review

"Nielsen Norman Group." *10 Heuristics for User Interface Design: Article by Jakob Nielsen.* Accessed August 05, 2017. https://www.nngroup.com/articles/ten-usability-heuristics/.

Chapter 43: User Testing

Krug, Steve. *Don't Make Me Think!: A Common Sense Approach to Web Usability.* Berkley: New Riders, 2017.

NASA. Accessed May 28, 2018. http://voyager.jpl.nasa.gov/.

Rubin, Jeffrey and Dana Chisnell. *Handbook of Usability Testing: How to Plan, Design, and Conduct Effective Tests.* Indianapolis, IN: Wiley Pub., 2008.

Chapter 44: Evaluation

Chamberlin, Edward Hastings. *The Theory of Monopolistic Competition ...* Cambridge & London, 1957.

Valen, Leigh Van. "The Red Queen Lives." *Nature* 260, no. 5552 (1976): 575. doi:10.1038/260575a0.

Index

A

Abandonment, 21, 185, 189, 262, 283, 290–291

Accessibility
  closed captioning, 120
  GOV.UK, 118–119, 121
  iOS Speak Screen narration, 119
  Microsoft Xbox, 120
  OXO Good Grips, 120
  resources for more information, 121
  sidewalk curb cuts, 120

Accommodation, 163

Act of volition, 150

Adams, John Quincy, 124

Adaptation, 162–163

Aeroxon, 175

Affordance, 45–46

Agile
  The Agile Manifesto, 213
  Agile UX, 213–216

Alcohol proof, 236

Alice's Adventures in Wonderland, 319

Allegory of the Cave, 270–271

Amazon Standard Identification Number
  (ASIN), 269

American Library of Congress, 266

2016 American presidential election, 190

Anchoring, 186–188, 192–193

Apparent motion, 82

Apple App Store, 176

Aristotle, 123–124

Arrangement
  deductive, 127–128
  inductive, 127–128

The Art of Persuasion, 124

Assimilation, 163

Attention, 89
  bit rate, 98
  span, 92–93
  sustained attention, 92–93

Attention blindness, See Inattention blindness

Audiences
  targeting, 40

Authority, 147

Automatic processing, 91

B

Baumeister, Roy, Dr., 150

Biases
  sampling, 237
  selection, 240

Birth traditions
  Bulgaria, 70
  Finland, 70

Book of Speed, 59

Book's origin, 67

Bottom-up processing, 73, 80–81
  perceptual errors, 81

Bouchard, Pierre-François-Xavier, 39

The Bridge on the River Kwai, 188

Britannic, 238

Bros, Warner, 43

Brunvoll, Grete, 186

Burj Khalifa, xiii

Business goals, 5–6, 260–261

C

Cacioppo, John, 153

Calculator Research, 236, 243–245

Carroll, Lewis, 319–320

Castaway, 299–300, 302–305

Causality, 238–239, 242

Cave paintings, xi

Chekhov's gun, 32

China's first emperor, 265

Chinese rail workers, 132, 323

Chinese sea snake, 132

Chinese state-sponsored commenters, xiii

Cialdini, Robert, 187, 274

Clark, Roy Peter, 261

Code Spaces hack, 267

Commercial Advertisement Loudness
    Mitigation (CALM Act), 79

Competitions, 14–15

Complexity, managed, 32

Confucianism, 196

Context, 22, 25, 37, 38, 204–205

Contextual inquiry
    behavioral observation, 251
    ethnography, 250

Contrasts, 186–188

Controlled processing, texting while
    driving, 91

Conversion rate, 237–238

Copernicus, Nicolaus
    heliocentric theory, 140

Correlation, 238–239, 242

Cowan, Nelson
    short-term memory chunks, 109

Crazy From the Heat, 262

Creating a process, 209

Crossing the Chasm, 45

Cryptolocker, 267

Cryptomnesia, 110

Csikszentmihalyi, Mihaly, 98, 192

Cues
    extrinsic, 172, 179
    intrinsic, 172, 179

Curse of knowledge, 44, 47

D

Dahl, Gary, 63–64

The Dating Game, 277, 280

Death rituals
    Yanomami tribe, 70
    Celestis, 70

Decision fatigue, 148–150

Decoy effects, ethical anchoring, 187–188

*The Demon-Haunted World: Science as a
    Candle in the Dark*, 76

Democratic People's Republic of Korea,
    See North Korea

The Design of Everyday Things, 45

Destructive operations, 188–190

Differentiation, 175–178

Disability, 117–118, 120–121

Discounts, 168

Documentation, 265–275

E

*The Ecological Approach to Visual
    Perception*, 45

Edge-cases, 279

Efficiency, 102

Eiseley, Loren, 140

Elaboration likelihood model (ELM), 153

Empathy, 135

Encyclopedia Britannica, 293

Enhydris chinensis, See Chinese sea snake

Ethos, University of Glasgow study, 125

Evaluation, 319–321

Exaggerated attributes and behaviors, 281

Experiences, deciphering, 42

Eyesight, 69

F

Facebook Beacon, 191

Facebook's Bullying Prevention
    Hub, 139

False consensus bias, 140–142

False data, 255–256

Familiarity, 44–46

The Fate of the Earth, 135

Feature creep, 159–160

Features, comparing, 29

Federal Communications Commission
    (FCC), 79

Federal Highway Administration, 162

Fidelity
    distracting, 272
    high, 272–274
    low, 272, 274
    necessary, 272

First Transcontinental Railroad, 131

Fishes'R'Us app, 11

Fitts' law, 204–205

The five whys, 256–257

Flow, 95

*Flow: The Psychology of Optimal
    Experience*, 98, 192

Fly catcher, 175

Fogg, BJ, Dr., 127

Forbidden fruit, 30, 32

Form fields, 31

Fort Julien, 39

Frederick the Great, 198

Free shipping, 24

Freud, Sigmund, 161

Frog, Michigan J., 43

Fusiform Face Area (FFA), 76

G

Gestalt Grouping
    proximity, 82
    similarity, 82–84

Get out of the building research
    (GOOB), 240

Gibson, James, 45

The Gift, 168–169

Girl Scout cookies, xiii

Gold coins, 95–97, 99

Good user experiences
    through delightfulness, 321
    through ease-of-use, 321
    through efficiency, 321

Google Play, 176

Gothelf, Jeff, 216

Great Wall of China, 166

Guinness, Sir Alec, 188

Gunpowder, 234

H

Hacking
    air-gapped data, 53
    Department of Veterans Affairs, 53
    target, 53

Help, providing to users, 21

Hercules, 1, 2

Hersey, John, 135

Heuristic review, scoring, 308, 310

Heuristics
    Gerhardt-Powals, 309–310
    Nielsen, 309–310
    Weinschenk and Barker, 309–310

Hiawatha train service, 85

Hick-Hyman's Law, 60

Hick's Law, *See* Hick-Hyman's Law

Hick, William, 60

Hieroglyphs, 40–42

Hiroshima, 135–136

HMS Hawke, 238

Hoagland, Richard C., 74

Hobson's choice, 241

Hodge, George, 233

Humor, 160–161

Hurricane Katrina, New Orleans' levee
system failure, 267

Hutcheson, Francis, 161

Hutzler 571, 221–222

Hydra, Lernean, 1

Hyperfocus, 98

I

IconFactory's Twitterrific, 176

Idea containers, 172–173, 179

Imperial Library of
Constantinople, 266

Inattention blindness
Harvard University study, 90

*Influence: The Psychology of
Persuasion*, 274

Interference, 106, 111

International Space Station, 75

Interviews, group, 256

Ivan the Fool, 181–182

Ivory-billed woodpecker, 171–172

J

Jessop, Violet, 238

John Walker's friction light, 268

Journey mapping
acquisition, 286–287, 289–290
awareness, 286, 289–290
conversion, 286–288, 290
retention, 286, 288–290
weak connections, 290–291

Juran, Joseph, 183

Just noticeable difference
(JND), 79, 197, 201

K

Kahneman, Daniel, 102, 187

Kairos, 127–128

Kano modeling
delight, 300–301, 303–306
erosion of delight, 305
features, 300–301, 304–306
user satisfaction, 300–301, 306

Kano, Noriaki, 300

Katamari Damacy
The Prince of All Cosmos, 35

Kirk, Captain James T., 311

Knowledge mapping, 293–297

Kulas, Neil, 269

Kuleshov effect, 203–204

Kuleshov, Lev, 204

L

Lærdal tunnel, 209–210

Law of Effect, 109–110

Law of Similarity, 84

Laziness, 101–102

Lean
Lean UX, 216–217, 219

*Lean UX: Applying Lean Principles to Improve
User Experience*, 216

Levitt, Theodore, 173

Library of Alexandria, 265

Ling-Ling and Hsing-Hsing, 165–166

Linux, 161

Literature and Latte's Scrivener, 177

Logos, 125–128

M

Mad Libs, 57–58

Manual on Uniform Traffic Control
Devices (MUTCD), 162

Maps, 272

Marathon du Médoc
drunk runners, 20, 24
relation to user journeys, 19

Mars, Cydonia region, 75

Maru, Kobayashi, 311

Mauss, Marcel, 166

Maus tanks, 27, 29

Mayan Codices, 266

McDonald's McResources, 243

Memory
    explicit, 106–107, 111
    implicit, 107, 111
    role of amygdala and
        hippocampus, 107
    short-term, 106, 109, 111

Mencius, 196

Mental Models, compared to
    schemas, 78

Microinteractions: Designing with Details, 192

Microsoft Excel, 65

Milgram Experiment, 147

Milton and Matilda, 166, 169

Milton, John, 73

Minecraft, 64

Minimum viable product
    (MVP), 212, 216–217

Mittelstand, 175, 178

Mitterrand, François, 113–114

M&Ms, 262

Mock-ups, 272–274

Model-building, 76

*The Monuments of Mars: A City on the Edge
    of Forever*, 74

Moore, Geoffrey A., 45

Morris worm, 53

Motivation, 153

Mountain goats, 101–102

MUTCD, *See* Manual on Uniform Traffic
    Control Devices (MUTCD)

Muthuvan people, 157

Myspace, 189

N

Naming, 268–270

National Assessment on Adult Literacy
    (NAAL), 278

Nazi book burnings, 266

Neelakurinji flower, 157–158

Neisser, Ulric, 90

Newsletters, 197

The New York Times website, 64

Nielsen Norman Group, 4

Nike, 247–248

Nixon, Richard, 165

Norman, Don, 4, 45

North Korea
    elections, 207

Nursing, 249

O

Office Space, 161

Olympic history, 13

One Froggy Evening, 43

Online ticket purchasing, 22

*On the Revolutions of the Heavenly
    Spheres*, 140–141

Ortolan bunting birds, 113

P

Pacific Islands, 247–248

Pac-Man, 95–97

Palmolive, Madge the manicurist, 3

Panda diplomacy, 166

Paradise Lost, 73

Pareto efficiency, 184

Pareto principle, *See* 80/20 rule

Pareto UX, 184–185

Pareto, Vilfredo, 182

Parker, Dorothy, 161

Patent medicine, 131

Pathos, 125–128

Pawpaw tree, 89–90

Perceived security
    through error messaging, 54
    through interaction design, 54

Perception, 73–88

Permissions, requesting, 23

Personas
 aspirational, 279–281
 demographic, 278, 280
 historical, 279–281

Persuasion, 71, 131–134, 153, 155, 171, 191, 199, 200

Persuasive Technology, 127

Peterson, Lloyd, 105

Peterson, Margaret, 105

Pet Rock, 63

Petty, Richard, 153

Piaget, Jean, 163

Place, 203–205

Plato, 270–271, 273

P51 Mustang, 76

Population, 234–238, 240, 242

Post-hoc
 fallacy, 114, 116
 rationalization, 114, 116

Potato
 history of, 198–199

Power law, 182

Predictability, 218

Price, 181–193

The Principle of Least Effort, 102

Principles of Marketing, 173

Problem statements
 how, 224–225
 what, 223, 225
 why, 223, 225

Process, 207–208

Procrustean bed, 240–241

Product, 171–179

Product availability
 auctions, 200
 exclusivity, 200–201
 limited time offers, 199–200
 special editions, 200

Product forms
 actual, 173–174, 179
 augmented, 173–174, 179
 core, 173–174, 179

Promotion, 195–201

Prototypes, 272–273

Psychological barriers
 avoiding pitfalls of, 86
 in social media, 86
 triggers, 86

The Psychology of Everyday Things, 45

Ptolemy V, 40

Puck Man, *See* Pac-Man

Pulling up Sprouts, 196

Q

Qualitative research, 247–258

Quantitative research, 233–242

Questions
 direct, 255–256, 258
 indirect, 255–256, 258
 leading, 251, 253–254, 257–258
 loaded, 254
 open-ended, 252–253, 257
 using silence, 254, 258

R

Rationalization, 113

Rawls, John, 142–143

Reagan, Ronald, 135

Realmac's Clear, 177–178

Reciprocation forms
 balanced, 167–168
 generalized, 167–168
 negative, 167–169

Reciprocity, 165–169

Recognition vs. recall, 106, 111

Reconciliation, 259–263

Reconsideration of goals, 163–164

Red Dawn, 293

The Red Queen Hypothesis, 320

Relevancy, 157–164

Reliability, 50–51

Rewards
 memory, 109–110
 restrictions, 109–110

Rhetoric, 123–125, 127–128

Rice cultivation, 195

Riedell Total Competition ice skates, 299

Rittel, Horst, 138

RMS Olympic, 238

RMS Titanic, 238

Road signage, 162

Rogers, Everett, 44

Role of UX, 5

Roth, David Lee, 262

Routes to persuasion
    Central, 153–154
    Peripheral, 153–154

Royal Navy, 233

80/20 rule, 183

S

Saffer, Dan, 192

Sagan, Carl, 76

Sahlins, Marshall, 167

Sailing ship, 233

Samoan patterns
    malu, 247–248
    pe'a, 247–248

Sample, 235–238

Sampling errors, 278

Satellites, 75–76

Scarcity, 198–201

Schell, Jonathan, 135

Schemas, 75–76, 78, 87

Schemata, 108

Search queries
    news, 227–228, 231
    technology, 227, 229, 231
    versus, 227, 230–231

Secondary users, 143

Seiden, Josh, 216

Security
    researchers, 53, 55
    through absence, 55

Selective attention, *See* Inattention blindness

Selective perception
    confirmation bias, 85
    halo effect, 85

Senate Chamber Candy Desk, xii

Serapeum, 265

Serial position effects
    primacy, 108
    recency, 108–109

Shell, G. Richard, 187

Sherman tanks, 28

Shopping carts, 149

Significance, 235–238

Signifiers, 46–47

Simplicity
    through absence, 30, 32
    through addition, 32
    through reduction, 31–32

The Simpsons
    Everything is OK Alarm, 127

Small multiples, 80, 87

Smell, 36, 69, 73, 81, 172

Snake oil, 132–133, 174

Software bugs, division-by-zero error, 50

Software Engineering, 50

Software trials, 168

Solutions
    accommodating, 15–16
    embracing, 15–16

Somerville, Ian, 50

Speed, 57–62

Spratly Islands, 283

Stability, 50–51

The Star Thrower, 140

Star Trek, 311

Statistics
    generalizability, 236–237, 242
    reliability, 236–237
    validity, 236–237

Stefanov, Stoyan, 59

Stone Age Economics, 167

Stone, Rosetta, 40–42, 265

Storytelling, 123–129

Stroop effect
cognitive interference, 92
Stroop effect test, 92
Stroop, J. Ridley, 92

Summer Olympic Games, 13

Survey of Public Participation in the Arts, 278

Symbology, 172

System 1, 102

System 2, 102

T

Tamakorogashi, 37

Taoism, 196

TAT-8 transatlantic cable, 51

Tattoos, 247–248, 250

Technology bell curve, 44

Ted Bates & Co., 3

Terms of Service Agreements, Carnegie
Mellon University study, 91

Tetrodotoxin, 10

Texas sharpshooter fallacy, 239–240

Thaler, Richard, 187

A Theory of Justice, 142

Thinking Fast & Slow, 102

Thorndike, Edward, 109

Thoughts on Laughter, 161

*Through the Looking-Glass, and What Alice
Found There*, 319–320

Tolstoy, Leo, 181

Top-down processing, 74–76

Torafugu, 9–10

Torajan people of Indonesia, xii

Total prestation, 167

Touch, 35, 36, 69, 79, 81, 87, 200, 237

Trigrams, 105

Tufte, Edward, 80

Tumblr Tumblebeast, 161

TV commercials, volume, 79

Tversky, Amos, 187

Twain, Mark, 160

U

UFO, 75–76

Underhill, Paco, 187, 205

Unfamiliar experiences, 91, 93

Usefulness, 65

User experience design (UXD), 4

User experience research (UXR), 4

User goals, 21–22, 24

User journey, 20, 24

User needs, 5–6, 64–65, 260–261

Users, 10–12

User testing, 311–317
confronting fears over, 316
qualitative, 313–314, 316
quantitative, 313–314, 316
remote, 314–315

USS Yorktown, 49–50

UXD, *See* User experience design (UXD)

UX Principles, 1, 2

UXR, *See* User experience research (UXR)

V

Valen, Leigh Van, 320

Vested interests, 11–12

Viking 1 Orbiter, 74

Viscardi, Henry, Dr., 118

W

Waterfall, 209–218

Webber, Melvin, 138

Weber's Law, 79

Weber, Ernst Heinrich, 79

Where users are, 25

Where users are going, 25

Where users were, 25

Why We Buy: The Science of Shopping, 205

Wicked problems, 138–140, 144

Wilderness wildfires, 31

Willpower, 150

Work hours, 208

Writing Tools: 50 Essential Strategies for Every Writer
  get the name of the dog, 261

X, Y, Z

Xolo, 259–260

Xoloitzcuintli, *See* Xolo

Printed in the United States
By Bookmasters